Warning Design:
A Research Prospective

Dedicated to the memory of Mena Haberfield

Warning Design:
A Research Prospective

BY

JUDY EDWORTHY

AND

AUSTIN ADAMS

University of Plymouth and University of New South Wales

Taylor & Francis
Publishers since 1798

UK Taylor & Francis Ltd, 1 Gunpowder Square, London EC4A 3DE
USA Taylor & Francis Inc., 1900 Frost Road, Suite 101, Bristol, PA 19007

British Library Cataloguing in Publication Data

A catalogue record for this book is available from the British Library
ISBN 07484-0090-7 (cloth)
ISBN 07484-0467-8 (paperback)

Library of Congress Cataloguing Publication Data are available

Cover design by Jim Wilkie

Typeset in Times 10/12pt by Santype International Limited, Salisbury, Wiltshire

Printed in Great Britain by T. J. Press (Padstow) Ltd, Cornwall

Contents

Preface and Acknowledgements

Over the last dozen or so years there has been a great deal of research and development work on the design and implementation of warnings. Both of us have been involved in this work for many years, one on the auditory warnings and one on the visual warnings side. Our involvement in this work, and our interest in developing specific warnings-related theory, has led to the development of this book.

We would both like to acknowledge some of the individuals who have supported us both professionally and personally in our research endeavours in this area, and whose inspirations have formed an important input into this book. Our thanks go in particular to Elizabeth Hellier, Chantal Laroche, Kathryn Momtahan, Roy Patterson, David Sless, Shirley Warland, Michael Wogalter and Harm Zwaga, as well as many of the other researchers in the whole area of warnings. Although our thanks go to these people, we should hasten to add that we take full responsibility for what is said and that not all of these good people will necessarily agree with all that we have said. We should also acknowledge those who have funded many of our research endeavours, and these include the Defence Research Agency, Farnborough, UK, the University of Plymouth, UK, the University of New South Wales, Australia and Standards Australia. We would also like to acknowledge the many publishers who have agreed to allow us to reprint their tables and figures in this volume.

Finally, we wish to acknowledge the support of our spouses, Chris Edworthy and Christina Vun, without whom the production of this book would have been virtually impossible. One of us, Judy Edworthy, would particularly like to acknowledge the support of her parents, Maureen and Sid Jones, who have over the last few years been asked to look after more grandchildren than they could ever have imagined, for longer than they could ever have imagined!

JUDY EDWORTHY
University of Plymouth

AUSTIN ADAMS
University of New South Wales

Foreword

Given the somewhat unusual nature of this book we felt that it would be ergonomically kind to the reader if we provided an introduction which was more of a summary, at least of the philosophical aims of the book, so the reader would know what to expect. In providing this we realise that there will be some potential readers who will decide, without progressing further, that the book is not for them, but that is the aim of this summary, and at least we will have saved those people some time.

The Table of Contents will indicate that we do attempt to summarise research concerning various types of warnings, but our main aim is to question the direction of much warnings-related research. A warning, we point out, is an artefact – something which does not exist in its own right, but exists to reflect, in some way, the total risk or hazard associated with the product or situation to which the warning refers. Such artefacts should, if they are properly designed, help the observer to decide whether or not to comply with the warning. We argue that whether or not the user complies depends not just on the warning, but on the whole situation, including the user's past experience, familiarity with the product, level of skill in using the equipment, as well as on the costs involved in complying, and so on. A major reason for noncompliance may be the user's judgement that the perceived costs of complying with the warning outweigh the perceived benefits. Sometimes this may objectively be so, sometimes not. We see the warning as providing one input into this decision-making process, and in providing this input we see the warning as having two roles. One is to draw attention to itself in some appropriate way and the other is to convey information the user needs, but may not already have, in order to carry out the appropriate cost/benefit analysis.

A direction in which warnings research is increasingly moving is in considering the broader context in which warnings are used, looking at such variables as experience and familiarity. At the same time, research is still being pursued avidly in the area of warning design, where the major goal is to elucidate the effects of a large range of variables on compliance rates. It would be unfortunate indeed if these two lines of research were viewed as being at odds with one another. Our preference is to pursue both, bridging the gap through appropriate calibration between the referent itself and the warning associated with it. What we hope to do in this book is to provide the framework for an appropriately broad approach to the issue of warning design – particularly with respect to the relationship between the warning and its referent, although we have dwelt largely on the former – in order that specifc warnings-related theory and modelling may develop. Theory at present seems either to be absent, or too general.

We argue that there are two types of components to a warning, whatever its modality. There are those which are iconic or attention getting and those which are informational. With warning labels, the iconic content consists of components such as the colour, specific signal words and so on and the information is provided by the addition of wording. With symbols, the iconic components are things such as colours and shapes of backgrounds as well as aspects of the symbol itself, but the symbol also aims to provide information. With nonverbal auditory warnings the iconic part is the acoustic structure, particularly the urgency, of the warning and the information is provided through the learned meaning of the warning. With spoken warnings, the iconic parts are the pitch and stress patterns – generally referred to as the prosodic elements – of the speech and the informational parts are the words themselves. The division of warning components into these two categories is not clear cut and there are numerous examples in each of the modalities where the distinction is blurred. We leave it to the reader to ponder on this distinction, but we believe that it is a necessary one to clarify research issues in the warnings area.

In this book we focus primarily on the iconic elements of warnings, and more specifically we pursue the concept of perceived urgency. This can loosely be defined as the alerting quality of the stimulus being observed, although we would not wish to propose this as a formal definition. What we need to know are the relative alerting qualities of iconic stimuli in order to provide at least a rudimentary mapping between the risks we are portraying, and the warnings which are portraying them. We argue that generally it is the iconic parts of a warning which convey these alerting qualities, but to some extent this can be done through the informational parts of the warnings as well.

We also present two very simple measures which might be used in warnings research; we have called them a compliance score and an effectiveness score. The naming of these two measures is somewhat arbitrary, but it helps us to distinguish between the compliance that is likely to occur anyway, without a warning necessarily being present, and the 'added value' that the warning brings. Although the effectiveness score is a likely source of a measure which could be used to decide whether or not to provide a warning for a particular referent, we argue that at present there are too many unexplored dimensions on the decision-making side of the equation to advocate such a use. Knowledge of the relationship between relevant variables and compliance could be used in designing warnings, but an important use in the research context would be in modelling the relationship between warning design and compliant behaviour. We go to some lengths to propose a method of selecting the range of a variable which should be used to permit meaningful comparisons and generalisations across studies, a requirement which must be met if there is to be any progress in the development of warnings-related theory. The reader interested only in the practical matter of warning design may find some of our arguments about the equivalence of dimensions to be unnecessarily esoteric, but we believe they need to be considered. If theory and practice in relation to warnings is to progress we must aim to reach a situation where we have methods and criteria which will allow us to examine two different warning-related stimuli experimentally, and from the results be able to say, within known parameters, if they will have the same effect on compliance behaviour. It is towards these ends that we feel research should be directed.

Finally, on a practical note, as we discuss auditory warning design at length in the book, we appreciate how useful it would be if the reader could hear some of the sounds we advocate. At the time of writing we hope, therefore, to set up an Internet site for audio demonstrations.

J.Edworthy@plymouth.ac.uk

A.Adams@unsw.edu.au

CHAPTER ONE

Setting the scene

1.1 Introduction and overview

Warnings come in a vast range of types. Throughout our daily lives we are continually bombarded by verbal warnings, sirens, bells, warning 'beeps' on our computers, warning labels on packages and machinery, warning signs at potentially dangerous locations – the list is endless. Although they differ by modality, by design, and by the way they are used, they all have one thing in common. They are artefacts produced by a designer in relation to a situation or product which has some associated level of risk* additional to that provided by – or which the user could be expected to bring to – the situation or product itself. They are designed to provide to someone exposed to that product or situation information additional to that which that person could reasonably be expected to possess. The designer (in the broadest sense of the word) is trying in some way to influence the behaviour of the recipient of the warning. This could mean preventing someone from doing something that he or she otherwise might have done, or it could mean getting he or she to do something that might otherwise have been omitted. The receiver of the warning then has the task of deciding whether the advantages in complying with the warning outweigh the costs of doing so. We must stress that many products or situations to which warnings apply incorporate various inherent cues to the type and severity of the risks involved in use or exposure. These cues are not warnings but are functions of those products or situations and therefore act as additional cues to the recipient in making his or her difficult decision as to whether to perform the protective behaviour stated or implied by the warning.

Consider for a moment that one has been working on a book for a number of weeks and is beginning to feel tired, and to some extent rather ill. The

* By risk we mean some global measure including the severity of the hazard, the likelihood of injury or other consequence, and any other pertinent features associated with the hazard itself. We are aware that other, more specific delineations apply but for our purposes such a definition is adequate.

chances are that if one continues to work in this way one will be more seriously ill in the relatively near future. Colloquially, we would say that our body is giving us 'warning' signals that we should slow down. We may perhaps be able to interpret these cues, possibly on the basis of our prior experience of such events. In the technical sense to be used throughout this book these cues are not warning signals – they are informative stimuli which are part of the event itself and which are not separable from it. An explicit warning signal occurs when a friend or doctor tells us we are working too hard and that we should take things more easily. In this instance the warning is separate both from the event itself and from the cues at the individual's disposal in interpreting that event. Similarly, a pilot flying low has available a number of cues, some based on previous experience, in assessing whether the risks of flying low under the prevailing conditions are excessive in the face of whatever benefits may apply. A warning occurs only when, for example, a light flashing on the central warning panel is illuminated or an auditory warning is heard. The warning is an artefact and has an existence separate from the event being warned about, although its existence depends entirely on that situation arising, or having the potential to arise. Thus, it is essential to differentiate between the warning itself and the event about which it is warning, for a number of reasons which will be touched upon in this chapter and elaborated throughout the book.

In research and development programmes which look at warnings and the situations in which they are used some researchers might be interested in whether specific compliant behaviour occurs, without being overly concerned as to the source of that compliance. Such researchers might be interested in variables unrelated to the warning, or for various reasons they might not be in a position to manipulate details of the warning. For example, in the use of medication we could be primarily concerned about whether the patient uses the medication properly under a variety of circumstances unrelated to the warning. We hope and assume that any warning information provided will increase proper usage, but our ultimate interest would be in the extent to which proper usage occurs. The source of the appropriate behaviour, whether from the person's experience of that situation, the cues within the situation itself, the behaviour of other people with respect to that situation, or, of course, the warning sign is, in such research, immaterial. In such research it is appropriate to look at the whole package – warning plus situation or product – rather than to look at them separately.

On the other hand, if we are a designer of warnings then we need to be able to differentiate between behaviour which occurs naturally in the relevant situation, without a warning necessarily being present, and the 'added value' that the warning might bring. We need to know the particular effect the warning will have so that we can then assess the relative effects of different warning variables on compliance. For such a person the distinction between the amount of compliance with and without the warning is crucial.

Another reason we might wish to differentiate between the warning and the situation it represents is that as researchers it is quite easy, if we follow a path in which warning and situation are inextricably bound together, to find ourselves having to account for the whole of human behaviour. There is endless theory on perceptual and cognitive processes, social behaviour, and on a plethora of other psychological factors, all of which could be pressed into service in the quest to understand behaviour in a situation in which a warning might be used. As researchers it is essential that specific theory and theoretical frameworks are developed to underpin empirical effort. In developing a theory applied to warning-related behaviour we must decide whether we are going to develop such a theory to be applied only to the change in behaviour induced by the warning, or whether we wish to account for the behaviour which occurs both with and without the warning. If the latter then we may indeed find ourselves trying to account for the whole of human behaviour. Whatever our position in this dilemma we would be well advised to be quite clear about the distinction between behaviour with and without the warning.

1.2 Warnings and their referents

It is useful to think of the situation or hazard to which a warning is referring as the referent of the warning. This is a term used both in warning symbols research (e.g. Easterby and Zwaga, 1984) and more recently in auditory warnings work (Edworthy and Stanton, 1995). The warning itself, which we have already argued is an artefact, can be thought of as a representation of the risk associated with the referent. There are two distinct aspects to that risk, both of which are represented by the warning, and both of which can inform the decision-making process that the observer goes through when deciding whether or not to comply with a warning.

1.2.1 The warning

Warnings are artefacts. They are representations of the situations to which they refer. Look at the warning in Figure 1.1.

The fact that such a warning sign could never exist demonstrates the fact that warnings and their referents should be differentiable. In this case the

Figure 1.1 An unlikely warning sign.

warning sign is self-referencing, because the referent is the sign itself. This of course is rather a facile example but it serves to illustrate the point. Most warnings serve two functions. These are the alerting function, which is somewhat abstract, being emotive, or motivational, or both, and the informing function, which is more explicit. In the example shown in Figure 1.1 the alerting aspect of the warning is provided by the signal word 'Warning' which represents some level of risk or hazard to the observer (the word 'Warning' has many dimensions associated with it which is something of a problem for this particular word). It has been shown empirically that different signal words produce different levels of arousal when read (for example, Wogalter and Silver, 1990). Thus they can be used as general indicators of the level of hazard associated with the situation or product. The word 'Warning' could be replaced by other words such as 'Caution' or 'Danger', which might represent slightly different levels of hazard. Such an alerting function could also be provided by the colour of the complete warning, or perhaps just by the colour of the word 'Warning'. On the other hand, those aspects of the warning which appear to have only an alerting function can also have an informational content. Thus the word 'Warning' appears to be merely an alerting word but it also gives additional information about how concerned we should be in relation to the risk we are to find out about.

We can also think of the alerting function as resulting from the whole gestalt or pattern of the warning on that product, or in that situation. The signal word, if any, or a symbol, and other factors such as the colour and layout of the warning, or the apparent urgency of the sound of an auditory warning, will all contribute. Also relevant, as will be discussed below, is our knowledge of the situation in which the warning occurs. Together the factors which have an alerting function can be seen as the iconic aspects of the warning. Such aspects act almost instantaneously and require little conscious information processing. Occasionally we may read the smaller print normally considered part of the warning and our anxiety level may rise as we realise that the product is much more dangerous than the initial iconic indicators implied, but generally one of the aims of a warning is to produce a rapid alerting response which is appropriate to the product or situation. Our point here is that the alerting function results from more than just the signal word. It results from the entire warning-in-context.

The other part of the warning is primarily informational. The 'Dangerous warning sign' does not give us very much information – it could, for example, tell us what to do to avoid injury from the dangerous sign. A warning which was more specific may impart information about how to handle a dangerous product, for example. Such a warning would be primarily informational in nature, but the phrasing or layout may also have some alerting function. 'Handle with care,' or even, 'Keep out of the reach of children' can to some extent embody an alerting function. Thus the warning as a whole has both alerting functions and informational ones and it is hard to separate these when the warning is seen as a whole. It is certainly misleading to suggest that the colour alone, or the signal word alone, does the alerting and that the words which follow do the informing.

Thus there are two distinct aspects to a warning – the alerting and the informing. Warnings of different types, including those occurring in different modalities, will usually contain different mixes of alerting and informational content. Indeed, different people dealing with warnings and their referents will interpret different parts of a warning in different ways. For example, a warning label may contain minimal iconic information – it may not even contain a signal word, and may be printed in monochrome – but may contain lots of information. A warning symbol may be largely alerting, but may also provide information about the risks and hazards involved. Here, individual differences will appear because some observers will know the meaning of the symbol, and can therefore provide their own information, whereas others may have to react to the explicitly alerting features of the warning. An auditory warning may be overwhelmingly alerting, containing no information at all beyond the fact that something has gone wrong. Furthermore, the warning may not itself be an accurate representation of the risk involved, in which case it is inappropriately alerting. This is certainly true in many situations where auditory warnings are typically used, which we shall come to later. On the other hand the meaning of an auditory warning may be known, in which case it also serves to provide information.

1.2.2 The referent

The third set of cues which will inform our behaviour, and may constitute the most important set, comes from the situation itself. Take for example a hazardous product such as weedkiller. It is very likely that the product will carry a warning label, which we will be able to divide into alerting and informational components. The colours and layout of the packaging may also indicate that the product is dangerous. Other important cues, however, come from our previous experience of other similar products, or perhaps from previous use of the same product, and possibly include the odour of the weedkiller, its colour, its viscosity, our knowledge of its chemistry and so on. We might even go so far as to try out a little on a favoured weed to gauge its strength. These natural cues to the level of risk will give us many indications as to how we should behave with respect to the product.

We now come to the most crucial aspect of warning compliance, which is that people do not always comply with warnings. Relevant research will be discussed later, but one of the most likely reasons why they do not comply is that the perceived benefits are not outweighed by the perceived costs of compliance. Warnings are usually used where there are risks, and in such situations there will be both benefits and costs involved in complying with the warning. The situation in which the warning occurs will be assessed using previous knowledge, natural cues from the situation or product, and information from the warning. It could also be influenced by the personality or mood of the recipient. Some of these features will encourage the recipient to comply, and others will deter him or her from doing so.

Aside from the influence of the warning itself, which is to be the main focus of this book, it is necessary also to look at the decision-making side of the equation. Psychological research uses many models of cost/benefit analysis, often known as utility models, which consider variables encouraging a person to carry out a particular type of behaviour. These models also consider other factors, known as cost variables, which encourage the person not to indulge in that particular behaviour. The relative weights of the variables with a tendency to increase or decrease the level of compliance will ultimately determine the person's decision as to whether or not to comply with a warning. There are many varieties of such models, for example the Subjective Expected Utility model, the Theory of Reasoned Action, various health belief models, and others, all of which can be adapted for use in the modelling of warning compliance. Some of these will be discussed later in this chapter.

In addition, there is a large amount of risk assessment literature in which the levels of risk associated with particular situations, products and so on can be assessed in some objective way, usually by experts within a particular field. These will also be considered in this chapter because they can feed into the practice of warning design and development. For the moment, however, we will turn to one of the principle elements in this book, which is the appropriate mapping of the iconic aspects of a warning to its referent.

1.3 Urgency mapping

For warning signals and labels it is often difficult to differentiate between the iconic and the informational components. For example, the word 'Warning' can function both as a signal word with an associated alerting function and as an informational word (telling the observer that he or she is observing a warning and that some additional relevant information will follow). Spoken warnings also have a prosodic element which may function in an iconic fashion, because they concern not what is spoken, but the way in which it is spoken. In the case of auditory warnings this differentiation is usually clearer in that the sound has an alerting function and also a precise meaning, which may be known to the recipient.

There is one particular abstract iconic feature of auditory warnings – their urgency – which is of practical importance in that the urgency of the warning should relate in some systematic way to the hazardousness or risk of the referent. Warnings can be said to be appropriately mapped when the rank ordering of the urgencies of the warnings with which referents are associated is positively correlated with the rank order of the urgencies or importance of those referents. Figure 1.2a demonstrates appropriate mapping. In practical situations it has been shown that such matching is, at present, generally poor, and more in keeping with Figure 1.2b. For example, applied work in hospitals has

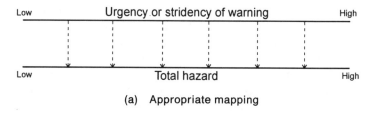

(a)   Appropriate mapping

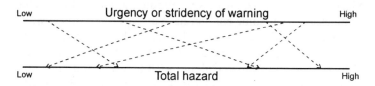

(b)   Inappropriate mapping

Figure 1.2   Urgency mapping (a) Appropriate (b) Inappropriate.

shown that the relationship between the situational urgency of the referent and the alarm which represents it is not correlated (Momtahan and Tansley, 1989; Finley and Cohen, 1991). That is, urgent-sounding alarms are often associated with non-urgent situations, and urgent situations may not have appropriately urgent-sounding alarms.

This is a particular problem because it has also been shown (Momtahan and Tansley, 1989; Momtahan *et al.*, 1993) that people working in particularly high-workload areas of hospitals (such as the operating room and the intensive care unit) are not good at identifying the meanings of alarms. If appropriate 'urgency mapping' were achieved – that is, if the apparent urgency of the warning itself were matched to the situational urgency of its referent, at least in relative terms within the warning set, then the recipient of the warning would at least know how quickly they should attend to the problem being signalled.

For a number of years studies have been carried out on the perceived urgency of auditory warning components (Edworthy *et al.*, 1991; Hellier *et al.*, 1993). The basic experimental paradigm has been to present subjects with relatively simple auditory stimuli, derived from different acoustic parameters such as speed and pitch, and to ask them to rate the perceived urgency. These studies have shown, as we would expect, that speeding up a stimulus makes it more urgent, as does raising its pitch or making it louder (Momtahan, 1990; Haas and Casali, 1995). The results of these experiments have been combined to create warnings which can be reliably differentiated from one another in terms of their urgency (Edworthy *et al.*, 1991).

From the results of this work on perceived urgency it has been possible to show the auditory warning designer how to manipulate the perceived urgency of auditory warnings. Thus the first step in achieving appropriate matching between warnings and their referents has been carried out. But would appropriate urgency mapping produce any improvements? There is a little research evidence on this too. Recent work (Haas and Casali, 1995) demonstrates that higher perceived urgency is correlated with faster reaction times, suggesting that faster reactions are obtainable from more urgent-sounding alarms (although we should guard against auditory warnings which are so urgent that they are aversive, a problem we will come to in later chapters).

Urgency mapping is also achievable with the iconic parts of a visual warning. For example, much research effort has been put into the elucidation of the effects of colour (Chapanis, 1994; Braun and Silver, 1995). This research shows quite clearly that some colours have stronger and others weaker effects on our assessment of the likely level of risk and hazard involved. Red always comes out as the colour associated with the highest level of danger, usually followed by orange and yellow. Colours such as white and blue tend to come further down. Researchers have also looked at the arousal strength of signal words (e.g. Wogalter and Silver, 1990; Wogalter and Silver, 1995). The signal words 'Deadly' and 'Lethal' tend to be highly rated in terms of the danger they connote, whereas words such as 'Note' and 'Notice' are rated much lower. Thus there is a general move in warnings research to examine the purely alerting aspects of warnings. The purpose to which these findings should be put, we believe, is in the appropriate matching of these alerting aspects to expertly derived levels of risk or danger actually present within the situation.

1.4   Warning measurement issues

Research is of little use unless its results can be generalised. To allow such generalisation it must be possible to compare studies which, for example, approach the same variables from different points of view. In the field of warnings research there are two methodological issues which restrict the opportunity for such comparisons. The first arises when studies do not include the control condition where there is no warning at all, thus preventing differentiation between the amount of compliance which occurs in a given situation with a warning present, and the amount of compliance which would have occurred anyway without the presence of the warning. The absence of this condition makes it impossible to compare the effects of the same critical variables across various studies because there is no baseline data enabling a comparison of the two situations. We will be proposing two types of score which can be derived from warnings studies to make such comparisons easier.

The second issue arises when attempting to compare studies when the range of particular variables which have been investigated is not the same. This becomes an issue when we want to draw conclusions about which variables have the greatest impact on compliance.

In this section we therefore deal with two main topics in relation to the assessment of the effectiveness of warnings. These largely concern methodological issues, and both relate to the importance of providing a distinction between the warning and its referent. The first issue is that of providing two sorts of compliance measures, one including the referent and the other excluding it. The second issue, which as we will see ties in with the first, is that of the calibration of variables and the selection of appropriate ranges for variables which might be included in warning compliance studies.

1.4.1 Measuring compliance

It has often been said that the ultimate test of a warning is whether or not it is effective (e.g. Peters, 1984). Most simply, this means that the presence of a warning should increase the amount of compliant behaviour above that which would be produced if the warning were not present. Some warnings can actually serve to reduce compliant behaviour – for example, there is nothing guaranteed more to make people look up than to provide a notice saying 'Don't look up!' It is useful therefore if we can see the 'added value' that a warning brings in terms of increasing the amount of compliant behaviour. The added value of a warning is usually shown by empirical studies involving a control condition which is usually the absence of the warning under consideration. This control condition can then be compared with various experimental conditions in which variables thought to affect compliance are manipulated, and their effects detailed. Such studies will be reviewed later in this book.

In some situations it makes little sense to look at the warning as a separate entity from its referent. Many factors will prevent us from carrying out studies where warnings which are deemed essential are taken away temporarily, and for many products the interaction between the product and the label are such that the warning label is seen as an integral part of the product. There may therefore be some circumstances where it is not possible to establish the baseline compliance level which would apply in the absence of the warning. Studies in relation to such products or situations are therefore limited to an examination of how compliant behaviour itself varies under a range of circumstances.

In order to clarify the situation, let us establish two simple scores which apply to behaviour in a warning-related study: an effectiveness score, and a compliance score. Let all such behaviour – both compliant and noncompliant – amount to 100 per cent (Figure 1.3). A certain percentage of this behaviour will be appropriate (compliant in this case) and will be exhibited without a warning being present. We can call this the baseline, and signify it as percentage 'a'. When we add a warning, compliance may increase to percentage 'b'. The difference between the baseline and percentage 'b' represents the degree to which the warning influences behaviour. We can call this the effectiveness score. It represents the 'added value' that the warning brings to the situation.

In some applications, however, we may want to know only about compliant behaviour under various circumstances, so we will not be particularly concerned about separating the warning from its referent. In this case we will

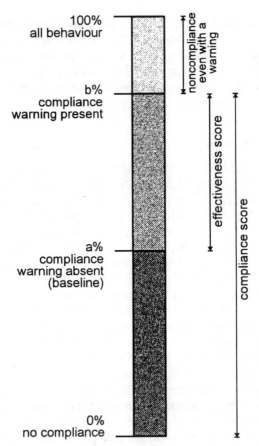

Figure 1.3   Compliance and effectiveness scores.

simply be interested in the percentage 'b' – we do not need to know (or may not be able to derive) the baseline. We can call this a compliance score. The difference between 'b' and 100 per cent is the extent to which noncompliance occurs in any given situation. The extent to which noncompliance is likely to occur is important in terms of the utilities attached to compliance and noncompliance, as we shall see later. Many experiments have been carried out in which the baseline has not been determined. The only sensible comparisons which can be made in these studies are on the basis of compliance scores.

Compliance scores are useful when we are carrying out field testing where, for example, if it were not for the presence of a warning we would not expect compliant behaviour to occur at all. Compliance scores are also useful in the direct comparison of warnings, but they will not help us as much as the effectiveness score in attempting to isolate the effects of the independent variables known to affect warning compliance. This is because they do not tell us the degree to which compliant behaviour was going to occur anyway, which may vary from study to study. Comparison of variables across studies may therefore be difficult.

Effectiveness and compliance scores would normally be measured in percentages calculated from observations of the numbers of people complying in a particular situation. Such scores could also be subjectively derived by asking people to estimate the number of people they think would comply in a given situation, or the degree to which they themselves would comply. For example, we may wish to observe the extent to which a 'Warning – do not walk under ladder' sign near a painter working up a ladder prevents people from walking under the ladder. If we wished to derive an effectiveness score we would first have to observe behaviour without the sign present. After establishing some measure of compliance (so that we knew when to count one person as 'compliant' and another as 'noncompliant') we might find that 50 per cent of the people observed demonstrated compliant behaviour. We would then add the sign and we might subsequently find that 80 per cent of the observed people complied. The effectiveness score for that particular warning would therefore be 30 per cent. In another situation we might carry out an experiment in which the baseline score (without a warning) is lower – indeed, many laboratory studies have found no compliance at all in control conditions, when they have been present – and where compliance rates shoot up to around 70 per cent when a warning is present. If the baseline compliance level (without the warning) was 10 per cent, then the effectiveness score in this situation would be 60 per cent and the compliance score would be 70 per cent. We propose these as two very simple measures which can be put to good use, and will be drawn upon in later chapters.

In a practical situation we may have two warnings which we may want to compare directly. We may set up an experiment in which some subterfuge is entered into, causing the subject to believe that he or she is participating in a task which has some risk attached. We then measure the degree to which compliant behaviour increases over some baseline control condition. If the difference is greater for one warning than for the other we can rightly conclude that one warning is more effective than the other. But we should consider the possibility that the warning which was the more effective of the two was possibly overcautious; that is, it was not an appropriate representation of the risk associated with its referent. The warning producing the lower level of compliance may have been the one more appropriately calibrated or matched to the situation than the other, seemingly more successful, warning. The missing link in Figure 1.3 is therefore the relationship between the level of risk inherent in the referent and the compliance level obtained (measured either as an effectiveness or a compliance score). In an ideal world there would be some general 'level of risk' scale which would map onto the degree of compliance obtained – with high levels of compliance being produced from high-value risks and low levels of compliance from low-level risks, and with appropriate calibration in between. A referent which produces, through its own cues, an appropriate level of compliance anyway is already appropriately matched. On the other hand, there may be other referents for which the addition of a warning – with both iconic and informational elements – may bring the level of compliance up to the appropriate level.

Warnings cannot be expected to produce complete compliance all of the time, no matter what their design, because we would then find ourselves spending a lot of time carrying out behaviour that would interfere too much with our daily lives. The ramifications of using a research and design approach which simply searches for the most effective types of design, and then uses or promotes the use of these ubiquitously, may lead to over-warning (in terms of warnings being generally too strident or fear inducing), with the prospect of eventual mass noncompliance. We should reserve the most 'effective' warnings for the most serious situations, and grade others correspondingly. After all, the warning is only one side of the equation. The other side, the cost/benefit analysis which the recipient is going to carry out anyway, will have a major influence on whether compliance occurs. In terms of Figure 1.3, the difference between b per cent and 100 per cent represents the extent to which noncompliance is the decision considered to have the greater utility. Making warnings more strident than they need be may not in the long term help the recipient make up his or her mind as to whether or not to comply. It may increase compliance for a particular application, but it may not help in the more general sense.

The most useful role for a measure of effectiveness is not in making decisions as to whether or not a warning should be used – the comparison of the 30 per cent from the ladder example and the 60 per cent from the laboratory study is to some extent meaningless, because there is not so much room for movement upwards with the ladder example – but in comparing the effects of different independent variables on compliance rates. In this regard there is a role for something akin to an effectiveness index, which may consider factors which would contribute to a decision on whether a warning is actually necessary. Such an effectiveness index would involve a weighted combination of variables such as the likelihood of injury, the possible severity, the type of people likely to confront the warning and so on, but the way in which these factors should be combined is not at all clear. At present, more general utility models may be of more use in making such a decision. We will come to these later.

If information about the effectiveness of various warning-related variables is available it can feed directly into the urgency mapping and calibration process. One requirement in that process is to create a range of warnings which vary in perceived urgency (or perceived severity, or some other measure of the importance of the warning). In creating such a series there would be little point in trying to use variables which we know can vary to a great extent without having much influence on the perceived urgency (or other measure) of the warning. Rather, we would use those variables which have a large effect when varied through a relatively small range. Those factors which do not have much effect on urgency might be used to create warnings which are different from one another, even though their urgency levels might be much the same. Of course we need to know a great deal more about how these variables interact, but there is a growing body of knowledge on the effects of individual variables.

At some point in this urgency-mapping research process we will want to make direct comparisons between variables, so it is important that they are comparable in some way. For example, if we want to know whether font size or border width has a greater effect on the perceived urgency of a warning label, or whether a change in pitch has a greater effect than a change in speed on the perceived urgency of an auditory warning, then we have to have some sensible rationale for choosing the range from which these variables are selected. Thus another important measurement issue is the determination of the range of a given variable which will be considered. It is this to which we turn in the following section.

1.4.2 Variable range

The range of variables already shown to have an influence on warning compliance is enormous, and other important influences undoubtedly remain to be discovered. Indeed, the charting of these effects is one of the major concerns of warnings research. In theoretical investigations we often need to know whether or not particular variables have a significant effect, no matter how small that effect might be. However, in applied work, such as that concerned with warning compliance and effectiveness, it is important to know the size of the effects and to be able to compare those sizes. One reason for this interest is that variables sometimes have effects which are significant but are so small that they are not of any practical use. Another and perhaps more important reason for our interest in effect size is that in the applied situation we would like to be able to tell a warning designer which of a variety of ways of altering a warning will have the greatest effects on compliance. As we shall see, we cannot make comparisons between the effects of different variables unless there is some rationale for equating the ranges over which the variables are to be compared.

The number of variables which have been examined for their effects on warning compliance probably numbers close to 100. Of course, only three or four such variables can normally be included in a fully factorial design. However, for our present purposes we can think of the empirical study of warning compliance as involving one huge multifactorial experiment in which all these variables are examined together. It is the task of this great experiment to find out which variables have the greatest effect, and to advise the designer accordingly.

If we were able to carry out such an experiment we would probably take some relatively simple measure of compliance. In practice, experiments in which large numbers of variables are examined can only involve a subjective measure, so we might ask a subject the extent to which he or she would comply if a given condition applied. Assume for a moment that we have analysed our data using some version of analysis of variance, multiple regression or perhaps factor analysis, and we have determined that some variables seem to account for more of the variance than others. We will be tempted to make the assumption that the variables which account for more variance are more important in warning compliance. If the differences in variance accounted for are very large we can be fairly sure that those inferences will be correct. However, if the differences are less than striking we have to be very sure indeed that we have compared like with like in terms of the range that each of the variables has covered. If, for example, one of the variables was examined over only a small part of its possible range, we could not legitimately compare its variation with that of a variable which had been examined over all its range. Let us demonstrate this with an example.

Imagine an experiment in which we are interested in the effects of changes in pitch (frequency) and changes in loudness (intensity) on the perceived urgency of auditory warnings. We want our results to inform the process of auditory warning design, so we really do want to know which of these two variables has the greater effect on perceived urgency. We therefore have to be sure that we are making legitimate and sensible comparisons between the two variables. There are a number of approaches to the comparison of ranges. Three of these are clearly possible in applied work, and each of them will be discussed in this chapter.

The first approach is to determine the full perceptual range for each variable and then to choose a proportion of that range. The proportion should be the same in each case. Luckily these two variables can be quantified, so this is not a difficult problem. When we look at the signal detection literature we find that the perceptual range of loudness runs from about 0–140 dB, and the perceptual range for frequency runs from about 20–15 000 Hz. Clearly we cannot cover these ranges in an experiment (nor would it be safe to do so) so we would choose some proportion of this range, say 20 per cent. Choosing the start points of the range is a further problem. We may carry out some pre-experimental studies in which we attempt to produce points of perceptual equivalence for these two variables, although the validity of any such method might be questionable. The absolute range that 20 per cent of these two ranges would cover is 28 dB in the case of intensity, and 3000 Hz in the case of frequency. The starting point might be 60 dB in the case of intensity and 100 Hz in the case of frequency. At first glance, the range for frequency seems very high but it has been derived by a logical method. It is likely that if variable ranges were derived in this way some interesting and surprising results might be obtained in warnings studies.

A slightly different version of this method would be to use the same number of Just Noticeable Differences (JND) in the ranges used for each variable, if these can be readily calculated. We can begin to see the kinds of problems which are going to arise if methods such as these are used. If we follow the JND route we run the risk of 'throwing the baby out with the bath water' because it may be that the reason one parameter has a greater effect than another is precisely because the perceptual range covers a larger number of JNDs than another. There are all kinds of philosophical and practical problems which will crop up if either of these routes is followed, even though they may make some logical sense. The second and third routes may provide a more useful solution.

At the other extreme, we might let the practical constraints placed upon the selection of variable ranges dictate our choice and range of levels. The most obvious point of reference here are standards. Thus, for example, if a survey of alarm standards shows that the intensity range over which an alarm can vary is 25 dB, and that the frequency range is 2000 Hz, then these might dictate the testable range for these variables. The argument behind this is that although there is no reason to suppose that these are perceptually equivalent in any way, they represent the range which is likely to be used. This is a completely practical route, and is one that often tends to be used in experimentation. There is a great deal of sense in this approach, but there is a problem with generalisation. If we use the limits set by standards as a way of setting variable limits then we will be doing just that – comparing standards recommendations. The ranges over which the variables may vary could well be quite different from the ranges recommended in standards.

The third possibility for defining variable limits is our preferred one. It is to use the ergonomic or usable range – or a fixed proportion of this range – for each of the variables under consideration. This range can usually be established through a combination of information from standards, perceptual data, guidelines and a working knowledge of the area. For example, the practical ergonomic range for an intensity variable is about 20–25 dB above masked threshold (assuming there is some noise in the environment) because once warnings are above about 15 dB above threshold they are hard to miss (Patterson, 1982) and so there is little point in increasing their intensity much beyond this. In addition, warning sounds become aversive if they are much above this upper limit, and so the use of such warnings would be unergonomic. For the frequency variable, sounds above about 1500 Hz can be aversive and so should not be used – even though current practice is to use such sounds. So in the experiment comparing intensity and frequency the 'ergonomic range' approach would be to allow the intensity variable to range from just audible to about 25 dB above masked threshold, and to allow the frequency variable to range from about 100 Hz (it would not be sensible to make warnings any lower in pitch than this) up to about 1500 Hz. In both cases the entire usable ergonomic range would be covered, and cross-variable comparisons would then be valid for the practical purposes to which they would be put. Although determining the ergonomically usable range for variables is somewhat tricky in some cases it has a number of advantages. The first is that the whole of the range can be covered in an experiment – if it can't be then one probably hasn't determined the proper ergonomic range. The second is that if this rationale is adopted there will be a rational basis for making cross-variable comparisons both within and between experiments. The third is that it provides data which can inform the process of warning design directly. Advantages which may be thrown out by a more theoretical rationale are kept – for example, if the ergonomic range of one variable covers more JNDs of a perceptual continuum than another, then so be it (we can take advantage of this during warning design) – but the approach is not limited by what actually occurs in practice. The primary principle is to examine the ranges that may be used in designing warnings, given that we want to develop warnings and warning systems which do not present the observer with aversive or extreme stimuli.

In many cases the application itself will determine the usable range of a variable. For example, the size of a warning label used in a study of warning labels could be dictated by the ultimate size of the warning label in practice, which in turn will be dictated by the size of the container on which the label will be placed. Many warning labels are placed on household cleaning fluid bottles which have a fairly uniform capacity, if not shape. This will dictate limits on things like the size of the font used, the size of the border around the signal word (if there is one) and so on. Of course, these absolute sizes will vary depending upon the product on which the warning is used – for example, a 'men at work' sign could, and should, be much bigger than a bleach warning label – but many of the questions asked about warning labels are more connected to proportions than to absolute sizes, so it is reasonable to generalise to some extent from warning labels studies if some sensible rationale has been used in the selection of variable ranges. A study in which this rationale was applied to the relative effects of font size, border width and white space on a warning label is described in Chapter 2. This study (Adams and Edworthy, 1995) also demonstrates how the apparent effects of variables can change depending upon whether one equates absolute values or comparable ergonomic ranges. Specifically, this study shows that one variable has greater effects in absolute terms on the perceived urgency of a warning label. However, the possible range for this variable is small. Another independent variable had a smaller effect in absolute terms but it could range over a greater set of values, thus it had a greater overall impact on the urgency of the warning. These kinds of effects, and the arguments which go with them, are important to know about if we are to generalise from warnings studies.

Of course, one of the greatest problems in any attempt to set ranges of levels is that many of the factors affecting warning compliance cannot be readily quantified. For example, it would be unwise to quantify colour because, although colours vary along the continuum of frequency, it makes no sense to quantify them because the frequency changes produce such striking qualitative differences. Other variables which might affect compliance may be even harder to quantify. But there are many which could be quantified if enough thought is given to such quantification. Many of the factors influencing the cost side of the cost/benefit equation can be quantified. For example, the distance one has to travel in order to comply can be easily quantified, and indeed this has been done (Wogalter *et al.*, 1989). The amount of money or time needed to comply can also be quantified. Many of the cost variables can be quantified on linear scales. It is the social variables that are probably hardest to quantify. For example, how does one quantify the variable of 'having a child with you and wanting to set a good example'? There are probably ways of taking pre-experimental subjective ratings of such variables, which could then assist the selection of ranges, but procedures for doing this would need to be carefully developed.

There are many issues yet to be resolved in the selection of appropriate variable ranges, but we believe that the 'ergonomic range' rationale is the best way to begin. Some studies implicitly adopt a logical approach to the problem, but we believe that an explicit approach is needed if we are to improve our ability to generalise from one study to another.

1.5 Behavioural issues

People do not always comply with warnings. Whether they do or not will depend on whether they judge the benefits of compliance to outweigh the costs. The benefits and costs may be actual, or they may be perceived and subjective, and possibly incorrect. Information assimilated from the warning – which may be accurate and objective or which may be more subjectively derived – helps the observer carry out the necessary cost/benefit analysis.

There are three main factors which need to be considered if we are fully to embrace the issue of warning compliance. If we are to map or match the warning's perceived urgency to some objective measure of risk associated with the product or situation in some sensible way then we need to master first the factors which affect the perceived urgency or importance of the warning, as we have discussed. These concepts are the primary concern of this book. Second, we need to master the process of risk assessment. This is an area in which a large amount of work has been carried out and which will be touched upon in section 1.6. The aim of risk assessment is to objectify information about different levels of risk inherent in different products, situations and so on. To keep our perspective on these two factors, however, we should realise that we need to know about them only to the extent that is necessary to determine an appropriate mapping between them, not because they are of any particular interest to us in their own right. As stand-alone variables they form part of the larger world of behavioural research and we are likely to be distracted from the cause of improving warning design if we focus on them separately.

The third factor, which we have so far only hinted at, is the subjective interpretation of the cues available in any situation in which risk is present (whether a warning is present or not). Again, our primary interest is not in this factor in its own right but as it relates to the objective level of risk involved. If this relationship is adrift it will affect the cost/benefit analysis that the observer makes, which in turn may lead to a decision to comply or not to comply which is not the decision with the greatest utility in that situation. Figure 1.4 illustrates the factors which we need to know about. One is the objective level of risk involved, as far as we are able to determine it. Another is the subjective perception of riskiness or hazardousness, which can take into account previous experience, perceptions of hazard severity, individual characteristics and the like which may vary over time. A third factor is the perceived urgency or importance of the warning, which derives primarily from the iconic alerting aspects of the warning. The iconic aspects, we argue, represent the direct route from warning to perceived urgency. The informational aspects of the warning, which are likely to be a representation of objective levels of risk (because this is information about the product or situation itself) contribute directly to the subjective judgement of the risk involved.

In Figure 1.4 there are two aspects which require mapping. The first is the relation between the iconic aspects of the warning and its perceived urgency. This relationship is primarily a function of properties of the warning itself. The second is the relationship between the objective risk and the subjective perception of that risk. This will be affected by prior knowledge of the product or situation but also by the informational properties of the warning. With an auditory warning the learned informational component will be the main one, but with a written warning the information derived directly from the warning, with no specific training other than the ability to read, will be an important

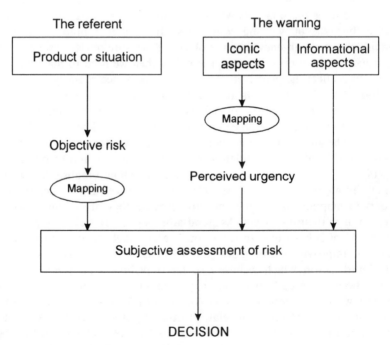

Figure 1.4 Components of warning compliance.

contributor. There are other relevant routes affecting the ultimate decision as to whether or not to obey the warning, but the main ones of interest are those shown in the figure.

If we think of the dimensions of warning compliance as falling into these three distinct areas, in the way shown in Figure 1.4, then we have a way of sorting out some of the more intractable problems about warning compliance. There is one issue in particular that will not be sorted out if we do not begin to think of the warning as an artefact or representation of the risk associated with its referent, as we are advocating in this book. There has been a fair amount of research impetus recently which has addressed the question of decision-making and utility models in compliance behaviour, where value expectancy models such as the Subjective Expected Utility model, the Theory of Reasoned Action, Health Belief models and others have to be considered in the understanding of compliance behaviour. In such models – which we will look at briefly in the next few pages – the warning is seen as part of the scenario which informs the decision as to whether or not to comply with the warning and carry out the necessary protective behaviour. For example, Dejoy (1991) argues that although there are many studies which show that factors such as familiarity, perceived product hazardousness, and costs affect compliance rates, the data on warning design are more ambiguous. He suggests that 'there is little data to suggest that even the best designed warning will override the beliefs and expectations that the individual brings to the situation'. The crucial point here is that, as Figure 1.4 suggests, warnings should not be expected to override such information. If that were their purpose they would be doomed to failure. Their role, we have argued, is to represent the risk as accurately as possible to inform the observer's decision. People are likely to be variously disposed towards carrying out whatever protective behaviour the warning advocates. These predispositions will be based on utility models which take into account the perceived costs and benefits associated with the behaviour. The purpose of the warning is to push this behaviour one way or the other by recalibrating the observer's decision so that, as far as possible, the objective and subjective levels of risk, and the associated costs and benefits, are matched and enshrined within the warning itself.

In this conceptualisation the warning is, again, separable from its referent. We believe that it is essential to do this to unpack the issues involved in protective behaviour. Utility models which have been used to look at protective behaviour tend to include warning dimensions among the many factors in the model. However, they cannot inform the process of warning design, which is to a large extent a separate issue. It is more useful, we believe, to view the warning as involving instantly alerting iconic aspects and possibly also informational aspects. The alerting iconic aspects should be an appropriately calibrated indication of the risk associated with the referent.

In the next few pages we will look briefly at subjective utility models and risk analysis approaches to clarify the role of warnings in relation to these ideas.

1.6 Utility models

When we receive a warning, we do so in a set of interrelated contexts. The
recipient, the warning and the situational context bring a number of variables
to bear which might affect compliance. The values at which the individual
variables are set can be used to inform the cost/benefit analysis which the
observer will perform either consciously or subconsciously. Some of these variables will have stronger effects than others. Indeed, it may be that some of the
more ephemeral situational elements have the strongest effects of all, although
currently we do not know the extent to which this might be the case. For
example, our mood at the time – which we may classify as a state variable –
might override all other considerations in self-protective behaviour. This does
not mean that the referent does not have the same level of risk associated with
it as it normally does. It does mean, however, that if we are ever to understand
self-protective behaviour fully we must look at these variables and incorporate
them into some predictive model of self-protective, or compliant, behaviour.

In order to do this, let us take as a starting point the Subjective Expected
Utility model (Edwards, 1954), a model widely used in attitude and decision-
making judgements in a number of spheres such as addictive behaviour
(Sutton, 1987) and more generally, health behaviour (Weinstein, 1993). Subjective Expected Utility theory proposes a relatively simple mathematical model
of decision making in which people assess the expected utility of an action,
choosing the action with the highest utility. The SEU of an action is the sum
of the perceived probability of each outcome multiplied by the desirability or
value of that outcome. In terms of warning compliance, the subjective utility of
complying with a warning might be seen in the following terms:

$$SEU_{comp} = P_{ncomp} \times SEV_{risk} - P_{comp} \times SEV_{risk} - COST$$

P is the probability of the outcome, which in this case is risk of injury, for
example (there are risks other than injury associated with warning noncompliance, however), so P_{ncomp} is the probability of injury if the warning is not
complied with, and P_{comp} is the risk of injury if the warning is complied with.
SEV_{risk} is a value corresponding to the risk inherent in the task.

Cost variables can include things like the amount of time or money
required to comply, and can include factors such as the distance one needs to
travel to comply. So, in words, deciding whether or not to comply with a
warning – or, more generally, deciding whether to demonstrate appropriate
protective behaviour in a situation with which risk might be attached – will be
associated with three main judgements. The first of these judgements is our
assessment of the probability that injury (or some other adverse consequence)
will result if the warning is not complied with (or appropriate protective
behaviour is not demonstrated). This is also related to the likely severity of the
injury. From this we need to subtract the probability that the injury will still
occur even if we comply (or show self-protective behaviour), again taking into
account severity. Thirdly, we also need to take away the costs associated with

compliance. We can reverse the first two parts of the equation to calculate the subjective expected utility of not complying with a warning (or showing self-protective behaviour), and then base our judgement about whether or not to comply on the basis of the figure which gives us the higher utility value.

Although the bare bones of the model are quite straightforward, it is necessary to develop submodels and to derive weightings for individual variables in the equation. Indeed, it is necessary to determine what those variables may be. Such models, if developed, would have to consider not only specific variable values – which can readily be derived from observation of the situation itself – but would have to be weighted according to the relative strengths of those variables on compliant behaviour. That is why it is essential to know what the relative strengths of relevant variables are, and to follow sensible rationales in the selection of the range over which those variables should be investigated, as discussed earlier. A crucial issue here, as alluded to earlier, is whether or not the design of the warning itself should be part of the equation. One can see arguments either way. Since we are interested in self-protective behaviour in circumstances in which a warning would typically be used we would expect the warning to be part of the scenario, so in this argument it should be included. The utility of the decision, however, should be based on the environ-ment not on the artefacts provided in order to shape behaviour. The warning should not influence the cost/benefit analysis, it should simply mirror it. There are some complex philosophical problems here, but on balance we feel that it is better to leave warning considerations out of the development of such models until the effects of other variables are clearly known. Keeping warning design issues separate from utility and decision-making issues is important if the research questions are to be clarified.

There are many other utility models, and derivatives of the SEU model, available for use in the study of compliance and other behaviours which are the result of some kind of decision making. These are generally models of social cognition, and include the Theory of Reasoned Action (Azjen and Fish-bein, 1980), which was later developed as the Theory of Planned Behaviour. There is also Protection Motivation Theory (Maddux and Rogers, 1983), Social Learning Theory (Bandura, 1977) and, perhaps most relevant of all, the Health Belief Model (HBM) (e.g. Schwarzer, 1992). Health Belief models have been used widely in the exploration of compliance rates in health care, focus-ing particularly on the extent to which the model can predict outcomes. The HBM has also been used in relation to specific types of safety behaviour such as seat belt use (Sutton and Eiser, 1990) and the use of bicycle safety helmets (Witte et al., 1993). The Health Belief Model, like many of its counterparts, focuses on the relative efficacy of numerous social, psychological and demo-graphic variables in predicting both actual and perceived compliance. It takes into account numerous hazard variables such as susceptibility and perceived severity, as well as the costs and benefits associated with the preventive action advocated. In all, they are likely to be useful in looking at many aspects of compliant behaviour. In terms of this book the most important question is

where such models sit in relation to warnings themselves. In the HBM, warnings could be incorporated into various elements of the model such as 'cues to
action', which include things like publicity and media coverage. Warnings elements can also influence the perception of risk and severity, which also usually
forms part of the model. It is again, however, our feeling that warnings should
be omitted and looked at separately because of the central issues of matching
and of appropriate calibration.

In conclusion, the important point about utility and decision-making
models is that they allow us to look at a range of social cognitive variables
which are likely to influence preventive action and self-protective behaviour. It
is beyond the scope of this book to look at these models in any detail – our
focus is on the design and evaluation of warnings, not on the origin of protective behaviour itself – but there are two critical points which should be
remembered with relation to these models.

The first is that they allow us to look at the subjective dimension of compliance, considering many of those variables likely to be important in warning
compliance. Generally they are more concerned with compliant behaviour
than they are with the effectiveness of warnings *per se*, because they usually
focus on behaviour which, when compliant, is advantageous to the recipient
(such as taking medication, participating in preventive medicine plans and so
on). The goal is typically 100 per cent compliance, with noncompliance being
seen as undesirable. The research aim appears to be to increase – or at least to
be able to predict – compliance rates.

The second issue concerns the relationship of these models to warnings
themselves. We are clearly advocating that design issues be separated from
utility issues, for a number of reasons. First, it is useful to keep them separate
because of the mapping argument. A warning cannot be mapped onto its referent if it is part of the referent itself. Secondly, design variables and social cognitive variables have different origins, and differential ability to change. For
example, one cannot really compare an attitude, which may have taken years
to develop, with a font size. Neither can one compare the presence of a red
background with demographic variables such as age and gender, which may
both affect compliance rates in different ways.

Although a long-term aim may be to produce models with weighted variables enabling the accurate prediction of compliance given a set of circumstances, the task at present appears to be more one of enumerating and
describing the effects of variables at an atomistic level which can be used grad-
ually to build up such a model. We believe that the warning and its referent
should be looked at separately. One way of addressing this division is to think
of the warning as servant to the referent, its aim being to present risk information – both iconic and informational – in as veridical a way as possible.

1.7 Risk analysis

The area of risk analysis is a vast one which is again largely beyond the scope of this book, but it has some relevance to behaviour in relation to warnings so we will touch upon it briefly. The central objective of risk analysis is to quantify, as objectively as possible, the total risk associated with specific situations and products. Risks can be assessed for situations in which the risks are fairly fixed and context independent, but they can also be assessed for situations which develop and change over time – a fire for example. There is a small risk that something like a small fire can develop into a very serious event. In such situations there is a tendency to use alarms when there is no risk, or a very small one, and hence a high false-alarm rate is likely. Hazard scenarios are often modelled using stochastic processes considering time progression, sensitivity and reliability measures, the expected number of people who may get injured and other processes (e.g. Pate-Cornell, 1986). Such models are very useful, especially as they can also help in determining the monetary costs and benefits of installing alarm systems. The arena in which risk analysis is placed is a large one, and the practical aim is to establish areas of risk and to reduce them as far as is possible. Warnings form but one aspect of this system, being called into play only when no further risk reduction is possible in relation to the referent itself. Warnings are never a substitute for appropriately good and safe equipment and product design.

The two main tasks of risk assessment are to assess exposure to hazards on the one hand and the effect of hazards on the other. Of course such calculations and assessments are multidimensional and are fraught with difficulty in practice. Total hazards also change over time. To carry out successful risk assessment the risk assessor needs access to a large amount of data which may or may not be available. If it is not available it may have to be approximated. The total hazard score for specific situations is the objective measure to which warnings should be matched. Essentially, a single measure is derived for a multidimensional stimulus to which we want to match a multidimensional warning which has a single urgency or stridency value. We wish, ideally, to achieve the kind of mapping shown in Figure 1.2a.

If the two measures – the objective measurement of risk and the subjective measurement of perceived urgency, or importance – are matched, we can have warnings which convey appropriate levels of urgency to the receiver. Within a particular work environment this may be done quite readily if the relative urgencies of both the referents and their accompanying warnings can be adequately defined. In practice, rank ordering of hazards and warnings may be as close as one can get to proper matching, but this in most circumstances should be good enough. Examples of how this might be done will be described later (Chapter 2 for warning signal words and Chapter 4 for auditory warnings).

One particularly useful concept in the area of risk analysis is the concept of a 'risk meter' (Brown, 1992). This rather tongue-in-cheek idea is a useful way of looking at risk analysis and calibration. Brown suggests that the idea of a risk meter is an attempt to emulate a hypothetical device (the meter) which could directly measure individuals and populations at risk with reference to specific materials and situations. Clearly such an idea is, and will remain, hypothetical but for our purposes it is the objective measure of risk against which an 'urgency meter' could be matched. The risk meter would calibrate the risk, and the urgency meter would generate the appropriately matched warning. It is the objective measure of the referent to which our appropriately calibrated warning is matched.

The 'risk meter' is in fact the idealised risk assessor. To get over some of the problems which risk assessors generally face, Brown proposes that the meter should have a fuzzy needle which defines only an area on the meter, and not exact values. In some cases the meter may need a jumpy needle to take account of sudden changes in risk (such as in a year where there might have been a nuclear power plant incident). One of the other useful things about the meter heuristic is that it contains an area marked 'acceptable risk', another marked 'caution' and a further area marked 'danger'. With a little further imagination we could easily build into this meter the concept of urgency mapping so that the meter could indicate to us the appropriate urgency for a warning to be used in the situation being measured.

We do not intend in this book to explore the area of risk assessment – mostly because it is a large topic which is considered in numerous books and hundreds of journal articles – our focus, rather, is on the design of warnings. One of our central concerns in warning design, however, is that of warning calibration. It will be useful for the reader to have the concept of the risk meter in mind when this is being discussed.

CHAPTER TWO

Warning labels

2.1 Introduction

Ensuring that a warning label or some other visual warning such as a sign in a public place is at least visible when appropriate might be one of a warning designer's lesser problems, but it is nevertheless something which must be attended to at some stage in the process of developing and presenting a warning. Unfortunately, making such a label conspicuous and attention getting is as much an art as a science, since factors related to the entire label need to be considered. A good textbook on graphic design and typography, such as Baird *et al.* (1993), will be more informative than an attempt to use any of the available behavioural research data. We would like to be able to give advice, such as to lay out the information in point form rather than in a running paragraph, and so on, but even on this minor point there is little hard data. Ley (1988) gives an excellent summary of the available data in the context of the labelling of pharmaceuticals.

As well as reviewing much of the recent warning label research, we wish in this chapter to cover a number of issues which seem to us to be central to the design process. One of these issues relates to the problem that warnings are not read at the time when the information contained within them is vital. This is a function of the fact that even though a warning label may always be present, its contents may only apply occasionally and so it is essential that the information is presented in a temporally appropriate way. A further issue relates to that of the calibration of warnings, an issue we discuss at length in this book. We maintain that warnings should be appropriately related in their urgency or stridency to the nature of the risk or hazard in their referent. There are several pieces of recent research which suggest, for example, that warning colours and signal words can be matched in some abstract way, and that different combinations of these can trade off against one another. We have argued in Chapter 1 that at least some aspect of a warning or its presentation is an icon which performs some kind of an alerting function. These iconic aspects of warnings are largely a function of the details of their presentation such as their colour, precise choice of signal words and their typography. It is our contention that these aspects of a visual warning sign or label can be matched to the situation where the warning applies in much the same way as has already been done for auditory warnings in some applications. We will come to this in later chapters.

There are some traditional technical problems with warning labels which we will attempt to clarify in this chapter. A warning label, as well as containing signal words and general words of caution, and even symbols (see Chapter 3), can also contain other information telling the reader about the risk, or about what he or she should or should not do. We regard these as informational components, separable from the iconic components. Research has sometimes been confused about which of these aspects is being investigated. Another issue relates to that of the selection of variables, and the comparison of results across experiments and the generalisation of results to real-life situations. This is also related to the problem of measuring effectiveness and compliance scores, which we will also examine. Many studies have measured simple effectiveness scores, and many of the studies which have not done this still allow the calculation of such scores, so these studies will also be reviewed.

2.2 Definitions

Weinstein *et al.* (1978) have pointed out that it is difficult to come up with a definition of a warning, particularly with reference to warning labels, because written warnings often contain instructions. Sometimes those instructions allow us to evaluate the importance of the warning, so it would be misleading to dismiss them as somehow separate from the main body of the warning. However, the informational aspects of written warnings are not normally those aspects which gain our attention and which enable us to make a quick judgement of the importance of the warning or of the extent to which we might feel it is necessary to absorb any associated informational content, although they might be. Consider, for example, a written warning with a section headed 'Safety Directions'. It is possible that we will see this and judge the warning which contains it to be more urgent in some way than a warning without that subheading. However, for the moment we will ignore such complications and concentrate on the more major attention-getting or urgency-imparting aspects of a warning, which we have termed the iconic aspects, and separate those from the mainly informational aspects. With both auditory and written warnings the alerting or urgency-imparting aspects can be quantified to some extent.

2.3 Legal aspects

Placing a warning on a product because it is likely to be effective is unfortunately not the only reason for such placement. A manufacturer may place a warning label on a product because it is mandated (even though it may actually be ineffective) or because this affords some protection against litigation. This second issue is a complex one and requires some discussion.

Peters (1984) points out that the legal origin of warning theory emanates, rather unexpectedly, from the Nuremburg War Trials. These trials, it is said, effectively internationalised the concept that each person must consent to his or her personal exposure to risk, and that such consent must be informed, voluntary and revocable. Hence the disclosure of safety information, which of course includes warnings, is a desirable safety objective. The most important point about this is that warnings should never be seen as a substitute for good design, and that designers should seek to eliminate danger and risk wherever possible. If injury does occur, prosecution can occur on a negligence or a strict liability basis, for which different logic applies. Under strict liability law it is the product itself which is at fault; thus a 'failure-to-warn' case should not go ahead on a strict liability case because the presence or absence of a warning is not at issue – it is the product itself which fails to reach the safety standards that it should. Weinstein *et al.* (1978) states that although 'failure-to-warn' cases can go ahead on other grounds, in practice such cases generally go ahead on the grounds of negligence. The general process of such cases is that injury or death has been sustained through the use of a product, or in some environment where risk is present, and the injured party or their representative is attempting to prove negligence through showing that warnings were either not present or were inadequate in some way. Strawbridge (1986) points out that a key concept in product liability cases is the 'but for' test. 'But for the presence of the warning, would the accident have occurred?' Juries, usually not experts in this area (although of course an expert witness may be called in), are thus required to make a binary decision as to whether or not the accident would have occurred if a warning had been present, or perhaps if a 'better' warning had been present about a complex multidimensional problem. Many of the problems likely to be encountered during such a decision-making process are those which provide interesting, and as yet still not fully explored, research questions. In inadequate warnings for example (in a case where a warning was present, but was deemed inadequate) what would be the features of a 'better' warning? Would it have a better physical design (in other words, would it be a better icon of the risk) in which case what features of the warning design would most improve its effectiveness? Did the warning not provide enough information? Was it the placement of the warning, or the size of its lettering, which reduced its effectiveness? Was it the familiarity that the user had with the product that resulted in the user's not reading the warning? Was the warning inappropriately mapped to the situation it was designed to signal? All these factors will feed into the decision-making process and many of them have been researched to some degree. The issue, however, is one of comparison and generalisation. How does even the expert, let alone the non-expert, compare the relative effects of these variables? We have argued for a system whereby variables can be equated in some way so that comparisons of this

kind can take place not only for the purposes of generalisation from experiments, but for direct application in legal cases.

Then there is the more typical problem of warnings simply not being present, where the case proceeds on the 'failure-to-warn' basis. One dimension of this problem is unrelated to warnings as such, but is related to the question of whether the equipment itself is badly designed and should be improved in terms of its safety. But if it can be established that the equipment is not at fault, then how is the jury to decide whether the presence of a warning would have averted the accident? In some failure-to-warn cases there is some environmental risk about which nothing can be done, such as a dangerous river, or unstable cliffs. How do juries decide in these cases? Again, many of the issues are at the heart of the research problems involved. For example, we have earlier advocated the use of simple scores for the measurement of warning effectiveness. The 'added value' of a warning can be calculated in some cases and used to demonstrate that appropriate behaviour takes place more often, or more readily, when a warning is present than when it is not. Thus we could argue that obvious risks, which are likely to lead to high levels of compliance anyway, do not need a warning whereas those which are less obvious may require some warning. Legally, this argument has been used often as a heuristic, and is known as the 'patent danger' test (Lehto and Miller, 1986). By this logic a risk which is obvious does not need to be warned about, but one which is not obvious (even though it may be less severe) might indeed need to be warned about. But how obvious does a risk need to be to require no warning? How obscure and unlikely does the risk need to be before a decision is taken not to provide a warning, even though the risk is non-obvious? If the risk is very severe (for example, if it is potentially life-threatening), then shouldn't it be warned about even if it is incredibly obvious? There are all kinds of problems which interfere with the proper functioning of this rule-of-thumb, not least of which is the fact that if we were to warn about every single risk in every situation, there would soon be a point where warnings might be less effective than they are often shown currently to be. A useful tool might be a method of calculating the utility of a warning label, and there have been some attempts to do this which will be considered later in this chapter.

Another issue here is the impact that the absence of a warning label may have on the use of a potentially risky piece of equipment. If the user perceives a product to be risky, looks for a warning label and then fails to find one, what is she or he to think about the safety of the product? To what extent has the ubiquitous use of warning labels become a substitute for our own good judgement about the safe use of equipment?

The legal situation provides a microcosm for many of the research questions still in the process of being answered. Many of the issues a jury might need to address in either a failure-to-warn or an inadequate warning case are related to the issues of effectiveness measurement, placement of the warning, proper calibration of the iconic aspects of the warning, and the comparison of the effects of different variables likely to influence effectiveness in a way that makes such comparisons valid. It is to these matters that we turn in the following sections.

2.4 Calibrating warnings

Recently within the warnings literature there has been an upsurge of interest in warning calibration, that is, in the psychophysical question of how physical variables associated with the design of warnings relate to psychological variables such as perceived intensity or urgency. Part of this interest is concerned with the extent to which there can be trade offs between various design variables in achieving any particular level of the relevant psychological variable.

For a written warning the situation or product itself will have an inherent level of risk associated with it, and it is the task of the reader of the written warning label (who is also, presumably, the observer of the situation or the user of the product) to make up his or her mind as to this level of risk. Models of behaviour and decision-making which could ultimately be used to assess this understanding were briefly reviewed in Chapter 1. With written warnings the major aspects which both gain attention and impart an immediate impression of the level of danger or risk are design features such as the signal word or words used (DANGER, WARNING, etc), and the colour and letter size used for those words. Other design features such as borders, white space around the signal word and more derivative features such as overall colour balance can also play a part in representing the urgency or severity of the situation being represented.

One of the prime arguments made in this book is that, at least within limited domains, the alerting aspects of warnings should be matched to the level of risk or danger associated with their referents – that is, that these should be matched at least in terms of their relative urgency. In the case of auditory warnings, the mapping of acoustic urgency to situational urgency is already beginning to result in design improvements. In the case of warning labels, relevant standards usually make an attempt to relate signal word or colour to urgency, but such relationships as are mandated are seldom based on the best available knowledge, although it must be said that until recently there has been little hard data that could be of use to the regulators. In what follows we will review both data and methods relevant to the urgency mapping of variables relevant to verbal labels.

2.4.1 Colour

Perhaps the most obvious iconic feature of a warning is its colour. This could be the colour of the entire warning sign, or of particular features within the warning. The vast majority of warning label research has used achromatic warning labels, but we might expect there to be some overall effect for colour in at least our subjective responses to warnings. A study by Kline *et al.* (1993), for example, required participants to rate various warnings along six different attributes which were 'salience', 'readability', 'hazardousness', 'carefulness', 'likelihood of injury' and 'familiarity'. The warnings were either multicoloured or were achromatic. It was found that along the two composite variables of 'perceived hazardousness' and 'perceived readability' the multicoloured warnings scored more highly than the achromatic warnings. This suggests that colour brings added salience to warnings in a general sense.

In addition to the obvious attraction of colour as a way of gaining attention and increasing salience (a feature widely used throughout the field of human factors), it appears that different colours are differentially associated with risk. In our culture, at least, the use of the colour red would signify a fairly high degree of risk and should thus serve well as an icon of fairly severe risks. A number of research studies have demonstrated that this intuition is well-founded. For example, Dunlap *et al.* (1986) have shown that participants rate the perceived hazardousness of colour words differentially, with the word 'red' being rated as associated with the greatest level of risk and the word 'white' with the lowest level of risk, the order in between being 'orange', 'yellow', 'blue', and 'green'. A study by Chapanis (1994) combined colours with signal words – a practice which has revealed some interesting results in other studies as well, to be reviewed later – and found that warnings presented in red produced the highest rankings, followed by orange, then by yellow, and finally by white. This finding thus almost replicates Dunlap *et al.*'s earlier study in which only the words associated with the colours were tested. Another recent study by Braun and Silver (1995) has also shown that the colour red is associated with the highest level of hazard, followed by orange, black, green and blue. So in general these results also show the same pattern. Particularly important in Braun and Silver's study is that in one part of the study actual levels of compliance were measured, and it was found that higher levels of compliance are produced when a warning is presented in red rather than in green or black although the results are a little inconclusive. A recent study by Adams and Edworthy (1995) also demonstrates that the colour red is associated with higher levels of perceived urgency than the colour black.

Thus the relatively small amount of literature on colour in warnings shows surprisingly consistent effects, although different colours have been studied in different experiments. The colour red is always rated as the one conveying the highest level of hazard or risk, and is commonly followed by orange and yellow. Lower levels of risk are associated with the colours green and blue, and white appears to be associated with the lowest level of risk. Theoretically, the reasons why this is so could prove to be interesting. For example, the response could be culturally determined or could be a physiological response to the colour itself which is independent of cultural influence. Whatever the source of this effect, the database so far obtained could be used to good effect in urgency mapping. Separating cultural from physiological effects would be an interesting research task and is one that has not yet been tackled, which is also true in auditory warning response (except for the cross-cultural work by Hoge *et al.* (1988), reviewed in Chapter 5).

2.4.2 Signal words

Many of the studies on colour have also included signal words, and signal words have also been investigated independently of colour. One of the most interesting findings here is that colour and signal words tend to interact, and to some extent trade off against one another so that, for example, the effect of a word associated with a high degree of risk can be weakened by assigning it a colour associated with low levels of risk. Similarly, the level of risk associated with a weak signal word can be increased by presenting the word in a strong colour such as red. Before we look at these effects we should examine the effects of signal words in themselves.

Signal words such as 'Warning', 'Dangerous', and 'Attention' may convey varying degrees of risk to the reader. In this sense they are very much like colour cues, and may serve as useful icons in the same way. A considerable amount of research effort has also been put into exploring the relative strengths of such signal words (for example, Bresnahan and Bryk, 1975; Leonard *et al.*, 1988; Wogalter and Silver, 1990; Kline *et al.*, 1993; Chapanis, 1994; Braun *et al.*, 1994; Braun and Silver, 1995; Wogalter and Silver, 1995). It is clear from the results of these studies that differences in signal strength between potential signal words are common and fairly robust. For example, Kline *et al.* (1993) presented three different signal words; 'Danger', 'Warning' and 'Caution' and found that the highest level of perceived hazard was found for 'Danger', followed by 'Caution', followed by 'Warning', although only the difference between 'Danger' and the other two signal words was found to be significant. Chapanis (1994) also found that 'Danger' was rated more highly than these two words, but the order between 'Caution' and 'Warning' was reversed.

Some studies have asked participants to scale the perceived hazardousness or arousal strength of a large range of signal words. For example, Wogalter and Silver (1990) asked participants to judge 84 potential signal words on six dimensions. These dimensions were strength, severity of injury, likelihood of injury, attention-gettingness, the level of care that one would show with a product so labelled, and understandability. All of the 84 signal words were rated along the six dimensions using an eight-point scale. The results showed that all the intercorrelations between the dimensions were significant, which led the investigators to propose that they were in fact measuring the arousal strength of the individual words – or at least a unitary dimension which is a general indicator of the perceived importance of the signal word. On the basis of their understandability, frequency of use and conciseness a shortlist of 20 terms was selected which covered the range of arousal strengths. This list is

Table 2.1 Twenty words ordered on arousal strength and six individual measures (from Wogalter and Silver, 1995)

Word	Arousal strength		Strength		Severity of injury		Likelihood of injury		Attention-getting		Carefulness		Understandability	
	M	SD	M	SD	M	SD	M	SD	M	SD	M	SD	M	SD
Note	2.12	1.90	2.46	2.03	1.61	1.97	1.64	1.45	2.57	1.93	2.32	2.07	5.50	2.12
Notice	2.80	1.87	2.89	1.59	2.39	1.99	2.75	1.90	2.96	2.03	3.00	1.80	5.25	1.65
Needed	3.16	1.92	3.29	2.03	2.86	1.69	2.82	1.79	3.25	1.55	3.57	2.40	5.86	1.74
Prevent	3.67	1.93	4.21	1.59	3.25	1.97	3.36	2.38	3.71	1.70	3.82	1.89	4.86	1.92
Careful	3.81	1.98	3.96	2.19	3.32	1.96	3.54	1.99	3.57	1.79	4.68	1.96	6.50	1.26
Alert	4.47	1.95	4.89	1.97	3.93	1.65	4.36	1.97	4.82	1.98	4.36	2.13	5.61	1.55
Alarm	5.19	1.75	5.61	1.71	4.64	1.99	5.04	1.69	5.32	1.59	5.36	1.75	5.96	1.64
Caution	5.26	1.82	5.32	1.85	4.79	1.75	5.50	1.67	5.32	1.76	5.39	2.06	6.14	1.43
Harmful	5.26	1.76	4.86	1.82	5.50	1.75	5.68	1.66	5.07	1.74	5.21	1.83	5.93	1.58
Warning	5.31	1.73	5.39	1.99	5.18	1.83	5.07	1.58	5.39	1.83	5.50	1.35	6.46	1.67
Beware	5.41	1.78	5.32	1.95	5.46	1.43	5.36	1.70	5.50	1.75	5.39	2.01	6.57	1.91
Urgent	5.41	1.84	6.00	1.83	4.93	1.82	4.21	2.35	5.86	1.63	6.04	1.43	5.43	1.55
Serious	5.51	1.72	5.46	1.44	6.04	1.62	5.50	1.75	4.79	1.81	5.79	1.93	6.07	1.58
Severe	5.55	1.70	5.89	1.59	5.68	1.39	5.75	1.92	4.93	1.74	5.50	1.79	4.25	1.69
Vital	5.80	1.84	6.36	1.39	5.68	1.76	5.43	2.28	5.86	1.82	5.68	1.81	4.18	1.59
Hazard	5.84	1.68	6.04	1.60	5.54	1.86	5.68	1.83	5.79	1.66	6.18	1.39	5.21	2.02
Danger	6.09	1.86	5.89	2.02	6.14	1.69	6.04	1.83	6.00	2.09	6.36	1.61	6.86	1.56
Poison	6.74	1.69	6.54	1.37	6.79	1.87	6.75	1.69	6.64	1.81	7.00	1.63	6.93	1.46
Fatal	7.20	1.44	7.54	0.69	7.39	1.57	7.07	1.30	7.04	1.35	6.96	1.97	5.75	2.01
Deadly	7.34	1.32	7.11	1.59	7.68	0.86	7.29	1.46	7.11	1.47	7.54	1.07	6.07	1.98

Key: M = mean arousal strength score; SD = standard deviation shown in Table 2.1. 'Deadly' is the term with the highest overall arousal strength, and 'Note' is the lowest. As well as being useful in themselves for warning label design, such lists of key words can be used in studies where trade offs are the main focus. Perhaps most interestingly, they can also be used to match actual hazards to signal words, which has also received some research attention (Leonard *et al.* 1989), which we will come to later.

As an example of the potential uses of this list of words, Wogalter and Silver's (1990) list of signal words has been used to explore the demographic stability of such terms, which is of practical importance. For example, Silver and Wogalter (1991) asked elementary and middle-school children to rate the perceived hazardousness of the eight key signal words explored in their earlier study. This study showed that although there were small differences between ages and specific signal word strengths, the patterns were similar for the children and the college students. This shows that the application of appropriate words to hazards – urgency mapping – could be an appropriate exercise for children as well as adults, because the connotations are understood from quite an early age. One result from this experiment was that the children produced higher ratings overall for the signal words, a finding for which many interpretations can be put forward, none of them as yet tested.

This work has been followed up more recently (Wogalter and Silver, 1995), in a study which has looked at the consistency of signal word ratings with the elderly, and with non-native English speakers, the latter being recruited from recent immigrants to the US. Across the two groups the same order of eight key terms was found, from 'Deadly', followed by 'Danger', through to 'Notice' and 'Note' at the other end. There were some small differences, but the interesting thing was that the order of the key words, the ones most likely to be used on products, was much the same. Correlations between the orders for children, the elderly, and college students were very high. Correlations were somewhat lower for those involving the non-native English speakers, although their results did correlate with the ratings for the elderly. Perhaps the most important difference was that the word 'Danger' produced a higher rating than the word 'Deadly' for the non-native speakers. Differences between the words 'Warning' and 'Caution' are confusing, as has been shown in other studies (Chapanis, 1994 and Kline *et al.*, 1993), thus their advocated use in many standards as a way of differentiating between hazards of varying strength comes into question. The confusion between the relative hazardousness connoted by the words 'Warning' and 'Caution' is such that, for example, in the Australian Standard on industrial warning signs the word 'Caution' has been removed altogether.

A useful practical outcome of Wogalter and Silver's (1995) study is a list of 15 words known by 95 per cent of fourth and fifth graders and 80 per cent of the non-native speakers, the two groups with the largest understandabilty problems. These are shown in Table 2.2. Although Wogalter and Silver (1995) point out that the criteria used to select these words are to some extent arbitrary (though logical), it does provide a useful information source for designers

Table 2.2 Mean carefulness ratings of signal words known by 95 per cent of fourth and fifth graders and 80 per cent of non-native English speakers (from Wogalter and Silver, 1995)

Word	Study 1		Study 2	
	Fourth and fifth graders	ASU college students	Elderly	Non-native English speakers
Notice	5.39	4.01	5.00	3.64
Careful	5.86	4.76	5.23	5.88
Alarm	6.16	5.01	6.09	4.87
Important	5.95	5.06	5.59	5.64
Caution	6.64	5.22	5.91	4.75
Don't	6.12	5.24	5.93	4.54
No	5.63	5.60	5.81	4.68
Serious	6.90	5.73	6.43	5.52
Never	6.09	5.93	6.27	5.34
Warning	6.52	6.13	6.49	5.58
Hot	6.00	6.21	6.61	4.40
Stop	6.11	6.43	6.95	6.55
Danger	7.12	6.49	7.00	7.63
Dangerous	7.18	6.64	7.04	7.66
Poison	7.49	7.00	7.57	7.93

of warning labels.

Wogalter and Silver's arousal strength list has also been used in the assessment of product labels. For example, Wogalter *et al.* (1992) have looked at the hazard ratings of realistic product labels using five words from their list (Lethal, Danger, Warning, Caution and Note) together with a warning symbol. They found that the presence of any signal word at all increased the perceived hazardousness of the product, and that there were additional differences between the extreme terms 'Lethal' and 'Danger' on the one hand and 'Note' on the other, although such differences did not exist for the intermediate terms such as 'Warning' and 'Caution'. No additional effects were attributable to the added symbol.

There is an issue relating to signal words which has to be considered, however. This is the point that the possible signal words have a range of different connotations, as well as demonstrating a general level of arousal. For example, some words which may be used such as 'Beware', 'Caution' and 'Careful' have action connotations. Others, such as 'Attention', 'Heed' and 'Note' have instructionally directed connotations, whilst others such as 'Risky', 'Urgent' and 'Deadly' have possibly more direct hazardousness connotations. Amongst Wogalter and Silver's (1995) list of signal words there are a wide variety of possible dimensions. Indeed, individual words can be placed on more than one of these dimensions.

One of the most problematic words is in fact the word 'Warning' itself. As well as having a (somewhat confusing) general level of arousal associated with it, it could have specific legal connotations. It could mean, for example, 'you will be fined if you do (or don't do) this'. It can also act as a pointer to information, telling you that the thing you are about to read is a warning. The word 'Caution', on the other hand, may imply no legal penalty, simply that you should show care when carrying out the task or activity.

Establishing which of the words actually relates most specifically to the hazardousness dimension may be a useful empirical study, especially if it is studied using careful scaling methodology and techniques. Parallel areas of inquiry can be seen for words which are used as quantifiers such as 'some' and 'many' (for example, Newstead and Collis, 1987) in which actual numerical scale values are assigned to quantifiers and are shown to have differing strengths, as we would expect. The use of intensifiers on the arousal strength of selected signal words (such as 'Extreme Caution', rather than simply 'Caution', for example) would also be of some interest.

2.4.3 Signal words and colour

Many of the studies looking at signal words have also included colour. Although no reason is generally given for the pairing of these two variables, we suspect it is because they both function as iconic features of a warning sign and are therefore implicitly bound together. There are a number of interesting questions which can be asked about colour/signal word pairings. First of all, do particular colours go 'best' with particular types of signal word? Furthermore, if colours with high hazard associations are paired with words of high arousal strength, are the effects additive? And even further, could such effects trade off against one another? To some extent each of these questions has been addressed, although the question of additivity (the extent to which the effects of individual variables can be added together to produce effects roughly equal to the sum of those individual variables) is possibly the most impenetrable of them. Let us look at these questions one at a time.

Chapanis (1994) asked participants to match the four colours tested in his experiment (yellow, orange, red and white) with the three signal words tested, which were 'Danger', 'Caution' and 'Warning'. Although the effects for 'Caution' and 'Warning' were inconclusive, participants showed almost unanimous agreement that the colour red should be associated with the word 'Danger'. This is not a surprising finding, but has a number of interesting ramifications. First of all, it looks as if participants are attempting to pair two icons (the colour and the signal word) so that they 'match' in some abstract way. Secondly, we might wish to know whether the effects are additive or whether they trade off, or indeed whether the effect lies somewhere between the two. In many ways this is a similar issue to that raised in the design of auditory warnings (for example, Edworthy et al. 1991; Hellier et al. 1993). It is well established that different sound parameters affect ratings of urgency, but what happens when sound parameters are combined and more complex warnings constructed? Is the resultant warning as urgent as the sum of its acoustic features (such as its pitch, its speed, its rhythm and so on) or is there a ceiling effect? Similarly, in visual warning construction, can colour and signal words be combined to produce yet more urgent warnings, or does some new urgency rating emerge? This is interesting theoretically, but it also has uses in warning design. For example, if effects are not additive, then there is little purpose in making a warning more complex using different signal words and colours. If they are additive, then a greater range of urgency levels can be created, which is good if such urgency-level distinctions are needed. The other side of this coin is that effects may trade off against one another such that two warnings could be made to be equally hazardous by using different signal-word/colour combinations.

Research results suggest that to some extent the effects are indeed additive. For example, Braun and Silver (1995) combined 21 of Silver and Wogalter's signal words (seven of which fell into a high arousal level category, seven into a moderate and seven into a low arousal category) with five warning colours (red, orange, green, blue and black), and asked participants to rate the level of perceived hazard. The signal word × colour interaction obtained is shown in Figure 2.1. This shows that words with high arousal strength are rated as less hazardous when they are shown in orange than in red, with this trend decreasing as the colours change through orange to green, blue, and then black. Figure 2.2 shows the specific effects for the individual signal words. Using this graph, trade offs can be seen which have obvious application. For example, 'Deadly' printed in green is rated as approximately as hazardous as 'Caution' printed in orange, and 'Danger' printed in red conveys approximately the same level of hazard as 'Attention' in red.

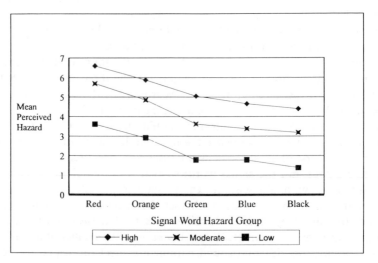

Figure 2.1   Interaction of colours and signal words (from Braun and Silver, 1995).

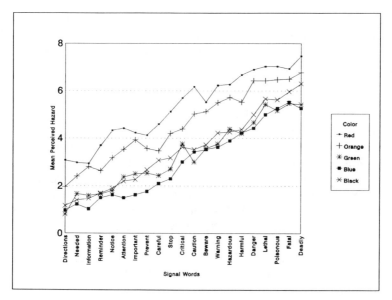

Figure 2.2 Mean perceived hazard for the combination of 21 signal words and 5 colours (from Braun and Silver, 1995).

Similar proposals have been made by Adams and Edworthy (1995) who explored trade offs between font size, border width, white space and colour. It was found that red warnings were perceived as being more urgent than equivalent monochromatic ones, and that the relative effects of font size and colour (the two most important variables emerging from the experiment) trade off to the extent that for the monochromatic word 'Warning' to be equally urgent to its red equivalent it needs to be about twice the size.

One important issue which needs to be addressed in this whole area of trade offs between colour and signal word (or some other warning design parameter such as font size, or border width) is its role in urgency mapping. On first sight, if the role of these essentially iconic features is to match the referent in some way, then such trade offs may serve only to cloud the issue. If we know that the word 'Danger' connotes a higher level of risk than the word 'Caution', how does it help the designer to know that the colour of the background, or of the signal word, can increase or decrease this perceived level of hazard? There are a number of dimensions to this question which will be discussed briefly below.

First of all, using a wider range of signal words and colours opens up a greater range of potential urgencies for the designer to call upon. Braun and Silver's study, for example, being a factorial design, produced 105 potential signal word/colour combinations which could be used in warning design, with the levels of relative perceived hazardousness of each combination clearly delineated.

Secondly, it means that warnings with approximately equal levels of perceived hazardousness can be produced by different means. In auditory warning design – where this has been done – it is the potential for confusion between warnings which is the driving force for advocating the design of warnings which are approximately equally urgent, and yet different from one another. In visual warning design it is the more practical aspects which are of concern. For example, if a designer wishes to design a warning with a fairly high level of urgency or perceived hazardousness, then it is very useful to know that instead of using red, which may not be feasible for a particular application, an equivalently urgent warning can be produced in monochrome by using letters which are twice as large. If, on the other hand, space is very limited then the designer may resort to using the colour red and reducing the letter size. Braun *et al.* (1994) have developed this concept further using an isoperformance technique which effectively creates lines of equivalent urgency for colour/signal word combinations. In Figure 2.2, points of equivalent urgency (say, a hazard level of or around 4 on the ordinate) can be read off, joined together and ascribed to a particular level of hazard or urgency. Several lines of equal urgency can be created, from the highest to the lowest levels. Each of the lines can be thought of as an isoperformance curve, because they represent points of equivalence in urgency or hazard rating. Examples of these curves can be seen in Figure 2.3, where the curves range from a hazard level of 2 to the most hazardous, level 6. The value of these curves is that the designer can adapt his or her design to the colours and words available, for a specific product or hazard, so that an appropriate level of urgency or hazardousness can be achieved. Of course, determining the absolute level of hazard is a diffi-

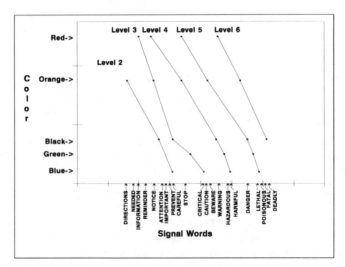

Figure 2.3   Isoperformance curves. Each curve connects values of colour and word that convey equal levels of perceived hazard (from Braun *et al.*, 1994).

cult problem (and is in the realm of risk analysis rather than warning design) but the issue of urgency calibration can be readily addressed by using such curves. The similarities between this work and work on auditory warning calibration are striking, as we shall see in later chapters.

2.4.4 Urgency mapping with signal words

We have seen that a great deal of effort is being put into the calibration of the iconic features of verbal warning design, particularly signal words and colour. The interest in such techniques is not, we believe, just the desire of researchers to know the relative effects of variables on warning design, but also a desire to produce databases which can be used in appropriate urgency or hazard mapping. That is, it should be possible to match warnings which indicate high levels of perceived hazard to those situations which actually represent a high level of hazard; warnings indicating a moderate hazard with hazards which are moderate; and warnings which indicate low hazardousness with low-level hazards. Many standards advocate this, but they often make assumptions about the perceived hazardousness of signal words and colours. The apparent misrepresentation of the words 'Warning' and 'Caution', with the latter generally proposed for lower level hazards than the former, is one such example of this misunderstanding. Wogalter and Silver (1995) suggests that the word 'Warning' is often used in inappropriate circumstances, suggesting that using the word in situations which do not warrant it might dilute its strength as a signal word. They go on to suggest ways of properly assigning hazard words to situations, one of which is to have groups of experts who are knowledgeable about the hazards and risks likely in particular situations rate these in terms of their degree of hazard, and then for some other person to assign appropriate words to the hazards. This procedure could easily be opened up by using Braun et al.'s (1994) isoperformance curves, and so could apply to more than just signal words, possibly even to all of the other iconic aspects of the warning. Such procedures have already been used in auditory warning design and implementation (see Chapter 4 for examples) and other studies in this area (for example, Momtahan and Tansley, 1989; Finley and Cohen, 1991) show that warning signals and their referents are often, in practice, inappropriately matched. We have no doubt that the same is true for the iconic features of written warnings.

If participants are given enough information about a situation, it appears that they are able to carry out matching of this kind. In a study by Leonard *et al.* (1989), participants were required to rank order 15 different types of risk on a 1 to 5 scale. These risks ranged from 'Driving the wrong way down a one-way street' to 'Failure to follow instructions for refrigerating or preserving food after opening the package', although many of the statements related directly to driving. The study found that participants were able to rank these risks consistently, and the rankings were fairly stable across age groups although there were some differences between participants who were older in age (in this case, above 25 years of age).

Subjects were asked in a second experiment to match five signal words – 'Deadly', 'Danger', 'Warning', 'Caution' and 'Attention' – to these risk scenarios. Previous work (Leonard *et al.*, 1988) had shown these signal words to be different in their arousal strength or perceived hazardousness, although the typical confusion between 'Warning' and 'Caution' had been found with their means being very close. The results of this matching showed that those signal words which were more arousing were, on the whole, matched to the more risky situations, and less arousing words were matched to lower rated risks.

This experiment shows that urgency mapping (although not under that name) can be done for signal words. Later we shall see that it can be achieved for auditory warnings as well.

2.5   Warning placement

Many of us hold an implicit belief that auditory warnings, if they are appropriately designed, are inclined to be more effective than visual versions of the same warnings. There is empirical evidence that spoken warnings for example are more effective than equivalent written warnings (Wogalter and Young, 1991; Wogalter *et al.*, 1993); this topic will be expanded in Chapter 6. If this is true (and the evidence is not incontrovertible) then it may well have something to do with the immediacy of the auditory channel, and the fact that auditory warnings are most commonly used in situations where a reaction is required in the near future. Nonverbal auditory warnings are by nature temporally proximal to the situation being signalled. Written warnings have a permanency which may bring benefits, but it may also have costs. The benefit is that they are always present, and no special technology is needed to trigger them when they are needed. Where the risk is unpredictable they may need to be permanently available. In many cases they may be appropriate at all times, such as a warning of dangerous cliffs. On the negative side, if a written warning is always present, and yet warns of a hazard which is appropriate only on some occasions (for example, if a product or piece of equipment is generally safe but can become dangerous if the user performs certain functions with it) the user has to decide when the warning applies and when it doesn't. The chances are therefore that the regular user will habituate to such warnings and may not understand their relevance under certain conditions.

Some years ago Cunitz (1981) drew our attention to a number of variables which may influence warning compliance, many of which have now been investigated in some detail and will be considered in later sections. However, two of these variables are particularly important here as they relate to temporal and spatial immediacy. Cunitz argues that a warning should be present when needed, with the implication that it should not be present when it is not needed. This variable he calls temporal immediacy. A second variable relates

to warnings being present where needed, which he refers to as spatial immediacy. Taken together these imply that written warnings should be close in
both time and space to the situation being warned about. Not only does this
suggest that written warnings should be made to behave, where possible, like
auditory warnings (which are usually temporally close to their referents), it
also delineates two variables which are clearly explorable in research. One of
the most attractive features of the dimensions of temporal and spatial proximity is that they are both represented by interval scales which can be measured
and calibrated quite accurately. The results of such experiments could fit nicely
into regression equations as well as the more typical analysis of variance
models. Such experiments have yet to be done using the regression approach.

A clear constraint on a pure research approach to these two variables is the
way in which products and equipment are used and the tasks which are performed when they are being used. For example, one may wish to calibrate a
spatial proximity variable on a linear scale with equal intervals, but find that
there are specific places around the equipment in which placement of a
warning would make sense, and others where it would not. Thus the experiments which have been done have tended to be driven by these practical constraints. An important feature of such experiments is that they focus heavily on
a task-analytic approach to warning design and placement. The task which the
user is going to perform is analysed in specific detail, and on the basis of this
analysis warnings can be placed at different spatial locations relative to the
task being performed – some near to the point where the warning is actually
needed, some further away. Temporal proximity tends to be fairly strongly
linked with spatial proximity in these studies, because a warning which is
placed centrally is likely to be observed at the time it is needed, as well as at
the place, but in general temporal proximity remains less explored as a variable than that of spatial proximity although it would be empirically easier to
impose control on the former. As Lehto and Papastavrou (1993) point out, we
would expect the limitations of short-term and working memory to have some
influence on warning compliance because it is well established that short-term
memory is very limited in its capacity (Miller, 1956). Many studies have
looked at the retention of warnings and warning information on the assumption that a warning which is better remembered is more likely to be complied
with. Temporal and/or spatial immediacy minimises the requirement for such
retention.

Let us examine those studies which have looked at warning placement, in
particular spatial proximity. Generally, these show that close spatial placement to the situation or specific task to which the warning is relevant has a
very marked effect on the level of warning compliance (for example, Frantz
and Rhoades, 1993; Wogalter et al., 1993; Dingus et al., 1993; Wogalter et al.,
1995). One result shown by several of these studies is that, when a warning
label actually interferes with the ongoing task, effectiveness increases quite
considerably. Wogalter et al. (1993) used the well-established chemical mixing
task (Wogalter et al., 1987) to explore a number of variables (such as visual

cluttering, which will be discussed later) including the embedding of warning information within the instructions for the task. The warning therefore interrupted the flow of the task. In another condition a much larger warning was presented, but was placed on a door which the participants could see while performing the task, and was thus visible throughout the task. Interestingly, the within-instructions warning was only 4 per cent of the size of the posted warning, but compliance levels were 33 per cent in the posted-warning condition and 92 per cent in the within-instructions condition where the warning actually interrupted the task. Thus a huge increase in compliance was observed in this experiment when the warning was temporally and spatially proximal. As well as a control condition, where there was no compliance at all, there was a condition in which both the posted and the within-instructions warning was present, and for which a slightly lower (but not significantly so) level of compliance was observed. This experiment therefore shows clear effects for both spatial and temporal proximity.

Another study showing a similar spatial/temporal proximity effect, as well as the more specific one of interrupting a task, is that of Dingus et al. (1993). As well as demonstrating the strong effects that cost appears to have on compliance levels, as shown previously in other studies (such as Wogalter et al., 1989), this study looked at different degrees of interactivity with the warning label. Three levels of this variable were investigated using a household spray bottle. These consisted of the usual type of noninteractive product labelling, a one-off type of interactive labelling known as 'billboard' labelling where the label had to be physically removed the first time the product is used, and a continuously interactive warning consisting of a trigger guard which had to be lifted before using the product and which made the warning continuously visible. In addition, during use the guard exerted pressure on the hand, serving as a constant reminder. As well as showing large effects for cost, this study showed that the billboard and the trigger guard conditions produced much higher levels of compliance than the normal label condition, particularly when the cost of compliance was low. The levels of compliance for the one-off billboard condition were much the same as the trigger guard condition, suggesting that continuous interference in the task does not result in further improvements in compliance above that produced by a more simple one-off label. Might this latter condition be likened to an auditory warning which, once activated, continues to sound throughout the task and is thus excessively interfering? Another feature of such warnings are that they are the type that users will try to inactivate if they possibly can, because they are excessively interfering. A final point here is that it would useful to know if the effect with the billboard label wears off over time because such labels are seen only once, the first time the product is used.

Another recent study which has shown large effects for temporal and spatial proximity (again, the two are inevitably confounded although studies which separate the two might provide interesting data) is one on filing cabinet use by Frantz and Rhoades (1993). Subjects were required to fill a filing cabinet but were warned, in one of four different conditions, of the usual problem with filing cabinets – that the cabinet might tip over if the top drawer is filled before the others, or has more material in it than the lower drawers. In the first condition the warning was simply placed on the box in which the new filing cabinet came, and so was not actually present on the cabinet itself. In the second, the warning was affixed to the bottom of the top file drawer. In the third condition it was affixed to the bottom of the top drawer so that it had to be physically broken in order to fill the top drawer. In the final condition the warning was placed inside the top drawer on a cardboard bridge which had to be physically removed in order to fill the top drawer. So in the third and fourth conditions the warning label actually interfered with the task, which was not the case in the other two conditions. The warning itself read

> WARNING. TO AVOID TIPPING YOUR FILE: (1) ALWAYS LOAD THE BOTTOM DRAWER FIRST, AND WHEN FULL, FILE IN THE NEXT DRAWER ABOVE. (2) ALWAYS LOAD THE TOP DRAWER AFTER ALL OTHER DRAWERS ARE FULL. (3) NEVER OPEN MORE THAN ONE DRAWER AT A TIME.

In all conditions but the first an identical warning was also placed inside the top drawer to satisfy the manufacturer's concerns for safety.

In all four conditions participants were required to unpack and arrange office furniture, including the filing cabinet, to which no undue attention was drawn. As the documents intended for the filing cabinet filled only half of one drawer there was no need for participants to use the top drawer at all, hence compliance to the warning could be measured. In addition, participants were also asked if they saw and read the warning. The results showed that the percentage of participants who noticed the safety information ranged from 0, in the condition in which the warning was on the box alone, to 93 per cent when the warning was in either of the conditions in which the warning actually interrupted task performance. The results also showed that the number of participants who read the warning ranged from 0 to 67 per cent across the four conditions. So of those who noticed the safety information, not all of them read it, a result also shown by Otsubo (1988) in an earlier study. Actual rates of compliance ranged from 13 to 73 per cent or 0 to 53 per cent, depending upon whether a lax or a strict interpretation of compliance was used. In all cases compliance was lowest in the condition where the warning was placed only on the shipping box, and highest in the conditions where the warning interfered in the task. For the objective compliance measures performance was slightly higher in the condition where the warning was on a cardboard bridge across the top drawer.

This study thus shows that spatial and temporal proximity have striking effects on the level of warning compliance achieved, and it again shows the particular effect that interference in a task can have on warning compliance.

A final experiment which will be briefly mentioned here is one carried out by Wogalter et al. (1995). This, too, demonstrates how the interfering property of a warning can influence the level of warning compliance. Subjects were required to install a computer disk drive. The location of the safety instructions, and the warnings which drew attention to this safety information, was investigated across seven different conditions. In one condition the safety instructions were inside the user manual. In a second they were also printed again at an earlier place in the manual (the second page). The other five conditions were identical to this second condition except that there was an accompanying label or leaflet which said 'Please read page 2 of the *Owner's Manual* before connecting equipment'. In these five conditions this label was placed on the packaging, on the cover page of the manual, as an accompanying leaflet, on the disk drive cable, and in front of the drive. The results are shown in Table 2.3. This shows that compliance increased across the seven conditions. The compliance measure used was a score from zero to three, counting if each of the three safety behaviours indicated by the warning was actually performed. The percentage compliance associated with each of these instructions is also shown in Table 2.3. Not all differences between conditions were significant, but the trend is very clear. One of the conditions required physical interaction with the warning (the front-of-drive condition) and this produced the highest levels of compliance overall, and was significantly higher than the basic and the redundant-warning conditions. The results also showed that less experienced users were more likely to comply, which is a result that has also been found in questionnaire studies (for example Godfrey *et al.*, 1983; Wogalter and Young, 1991).

The temporal and spatial proximity variables are only two of an extensive range of variables known to affect warning compliance, but they are ones that we have focused on here because they are increasingly becoming a topic of investigation in the research literature and are variables for which large effects on warning compliance have been repeatedly shown (and interestingly they are ones for which effects are unlikely to be revealed from questionnaire and other subjective studies). These variables are also of great practical importance. Investigations into their effects are largely driven by the task-analytic approach to warning design and implementation advocated by, for example, Lehto and his colleagues (for example, Lehto and Miller, 1986).

It is important, however, for research to extend these findings, using the regression approach, to give functions for both spatial and temporal proximity which would enable these two variables to be used in a general model of urgency mapping. For example, if high spatial and/or temporal proximity does have large effects on compliance, then this might be reserved for the most hazardous products. Trade offs could also be identified, such as any which might exist between warning size and temporal/spatial proximity hinted at in the Wogalter *et al.* (1993) study discussed earlier. Such work remains to be done.

The background to temporal and spatial effects is largely unexplored and there are many issues which require further elaboration. For example, it would be important at least from an experimental point of view to attempt to dissociate spatial from temporal proximity because the two are often linked – a

Table 2.3 Percentage compliance as a function of warning placement on a disk drive (from Wogalter et al., 1995)

Conditions	Mean compliance score (0–3)	Turned off computer (%)	Grounded plug (%)	Ejected disk (%)
(a) Basic manual	1.08	41.67	33.33	33.33
(b) Redundant-warning only	1.58	58.33	50.00	50.00
Redundant warning manual plus the supplemental directive on the:				
(c) Box	2.25	83.33	66.67	75.00
(d) Cover page	2.25	75.00	75.00	75.00
(e) Leaflet	2.50	83.33	83.33	83.33
(f) Drive cable	2.50	83.33	83.33	83.33
(g) Front of drive	2.75	91.67	100.00	83.33

warning which is in close spatial proximity to the user during a task is usually also within close temporal proximity. The two also need to be dissociated if we really want to make cross-variable comparisons, a point for discussion later. Adaptations of already-tested methodologies may prove useful in further studies. For example, in the Wogalter et al. (1993) study, participants could be told to look at the posted warning at a particular point in the instructions, rendering the posted warning temporally closer but further away spatially than the warning embedded within the instructions. In the disk drive study discussed earlier (Wogalter et al., 1995) this has been indeed been done, but the dissociation between the temporal and the spatial elements in the design were not specifically examined. A further aspect requiring elaboration is the particular effect that task interference appears to have. It would be interesting to know whether this is due to very high spatial and temporal proximity per se, or the interrupting quality of this interference. If the levels of spatial and/or temporal proximity tested could be systematically varied in some sensible way it would be possible to see if there is any 'added value' linked to the presence of interference over and above that of very high temporal/spatial proximity, and a range of techniques including regression could show this. The experiments which have been carried out to date do seem to show that the effect is greater than would be predicted by high proximity alone, and that there is a specific effect of task interference.

One of the implications of the results to date is that they suggest that warning labels would be more effective if they were to demonstrate the temporal and spatial proximity generally reserved for both verbal and nonverbal auditory warnings. What, then, of auditory warnings which are made to demonstrate the usual characteristics (at least in time and space) of written warnings? This question suggests more appropriate ways of making direct and meaningful comparisons between warnings in different modalities. The studies by Wogalter et al. (1993) described above, for example, included voice warnings which produced greater levels of compliance than equivalent written warnings, even when the voice warning was initiated at some point before the chemical mixing task began, rendering the voice warning non-proximal to the task. Although many interesting and useful variables were investigated in this study (to which we will be returning later) there was no condition in which temporal and spatial proximity variables were included together with warnings of different modalities so that direct comparisons could be made. Future experimentation might lead to such fruitful cross-modality comparisons.

2.6 Compliance and effectiveness scores

Sometimes it is useful to know about the degree to which compliant behaviour is exhibited in a particular situation, regardless of whether a warning is present or absent, because we are interested in safety in general. We proposed in the first chapter that it is therefore useful to elicit compliance scores for particular scenarios and situations, and possibly for different demographic subgroups such as men and women. These scores, ranging from 0–100 per cent, tell us about the behaviour and the warning together. They do not tell us anything about the 'added value' of the warning *per se*, but then we do not always need to know about this – and in some cases we cannot dissociate the two either for philosophical or ethical reasons. Many of the studies to be reviewed in the following pages measure the overall compliance level in this way.

In other situations we may want to know the added value that the warning itself brings to the situation, in which case we are more interested in effectiveness scores. These can be obtained by establishing baseline performance (behaviour when a warning is not present) and then subtracting this baseline from compliance levels after a warning is introduced. In some cases the baseline is zero, so the effectiveness score equals the compliance score. In other situations compliance levels might be fairly high anyway, so in order to make meaningful comparisons across variables, and to draw sensible conclusions about the effects that different variables exert on compliance levels, we also need to establish the degree to which compliant behaviour occurs anyway, without a warning being present. Many studies have also done this, although they have not always reported effectiveness scores explicitly.

Effectiveness scores might possibly be used to gauge the utility of a warning. For example, if it can be shown that the addition of a warning increases compliant behaviour by only 5 per cent, then we might not wish to incorporate a warning at all if there are costs in doing so. However, there are so many factors which would need to be taken into account in taking such a decision – any model would have to be weighted to consider aspects such as the likely population to which the warning was addressed, their familiarity with the product, the likelihood of injury, the severity of any likely injuries and the cost of compliance – that we judge it to be premature to propose any such use of effectiveness scores in generating utility scores. We feel that the utility question needs to be approached from a more general perspective, and one such approach will be considered shortly.

In this chapter we will not grapple with the difficult question of how the proposed compliance scores might be used to make decisions about whether or not a warning should be used in a given situation. At the present state of development in the study of warnings we believe that the best use of such scores is to clarify empirical issues such as those involving the comparison of effects across variables.

2.6.1 Warning utility

There is a case to answer concerning the overuse of warnings. Intuitively we believe it to be a problem but there is little empirical evidence showing how that problem might be manifest. If the problem can be demonstrated, such overuse may well be improved by proper calibration, as we are advocating throughout this book. If we wish to decide whether a warning should be present we may want to look at the use of warnings on a much more general, population-wide basis.

The possible problems that may arise through warning overuse are fore-shadowed by the study of Rothstein (1985) who found that the probability of recall of a given warning decreased as the number of warnings presented increased. In addition, warnings which warn about very minor risks, or risks which are extremely unlikely to actually happen, seem intuitively to reduce the believability of warnings which do actually warn of real risk. For example, Wogalter and Silver (1995) suggest that the use of the word 'Warning' in situations such as 'Warning: Discount coupon will expire at year's end' may dilute the arousal strength of the word if this is done too often. In a specific warning situation, it has been shown that clutter of a general sort lessens the impact of a warning (Wogalter et al., 1993). More specifically still, Godfrey et al. (1991) found that the amount of 'label clutter' significantly increased the time it took to locate the target warnings. Although we have little direct evidence, it would appear that visual clutter in the form of unnecessary warnings and information (which also take time to read) might also lessen the impact of warning labels, suggesting the need for some method of cutting down the number of warnings generally in the environment.

McCarthy et al. (1995) point out that most of the research on warnings, as well as most of the published standards on warning use, tend to focus on how to warn (that is, the content and structure of the warning message) rather than on what to warn about. Indeed, the implication of such standards seems to be that all situations and products warrant the inclusion of warnings. This may be the safe step in terms of a manufacturer's wish to protect against litigation, but it may not produce the best compliance levels in the long run. The authors go on to propose impact criteria for warnings using accident data as their basis. They suggest examining a variety of sources of information which are likely to reveal those situations, pieces of equipment and modes of use which are likely to produce the most frequent and severe accidents and for which warnings are therefore likely to be appropriate. For example, in earlier work McCarthy et al. (1982) proposed that warnings should be used on home products if they are associated with at least 1 per cent of the total risk for that area. Additionally, only those accident modes which account for at least 10 per cent of the total risk for that product should be warned about. Thus warnings about risks which could result in small or infrequent losses are eliminated. The authors' general procedure is therefore first to identify the products with the greatest overall risk, and then to determine which of those products have the most incidence of specific risks. A risk hierarchy can then be developed which can be worked down and decisions made about the appropriateness of a warning for each item on the list. The details of how decisions are made are not thoroughly fleshed out. For example some risks are so obvious that they require no warning. McCarthy et al. (1995) suggest for example that warning the user that scissors can cut would be fruitless. They also suggest that some index of the severity of the risk should be considered, but they do not make specific proposals as to how any such index may be calculated.

Another problem area lies in how to deal with items of equipment which are very heavily used and produce large absolute numbers of accidents but, given the heavy usage, are generally seen not to be (and in fact are not) risky products. In the home, for example, chairs are responsible for a large number of accidents because of their high usage, but they are generally used safely so warnings might be confusing and ineffective. The most useful data to have here would be the ratio of accidents to usage level, but monitoring usage levels and deriving such ratios would be impossible in many cases.

Approaches such as McCarthy *et al.*'s (1995) can tell us where the major risks lie, and can produce between- and within-product risk hierarchies, but how a decision should be made as to whether to provide a warning for a specific situation or product is still an open question, and one which requires further research effort.

2.6.2   Studies of warning compliance

2.6.2.1   Experimental studies

There have been a large number of studies on warning compliance, some of which have taken direct, objective measures of warning compliance while others have taken objective measures of variables thought to be correlated with warning compliance. Yet others have taken subjective measures in a variety of different ways, sometimes taking subjective measures of variables thought to correlate with warning compliance, sometimes taking single judgements of composite stimuli using regression models in order to evaluate the relative contributions of different variables to the overall judgement being made. Table 2.4 provides a simple summary of the main types of studies which have looked at warning compliance.

There are two main dimensions to this classification. The first is whether the study takes objective or subjective measures of compliance, and the second is whether compliance is measured directly or indirectly. A direct objective measure of compliance is one obtained by observing whether subjects comply or otherwise with a warning. A direct subjective measure is obtained by asking subjects to rate in some way if they would comply with a warning. An indirect objective measure is one where variables other than actual compliance are measured. In studies of this last type the dependent variables under consideration are thought to correlate with warning compliance but are not a direct measure of that variable. Variables here would include reaction time and eye movements. An indirect subjective measure is one where ratings are sought on dimensions thought to influence warning compliance, such as font size and

Table 2.4   Varieties of warning study

Direct objective measures	Laboratory experiments	e.g. Fake chemistry task, saw use
	Field studies	e.g. Broken door, slippery floor
Indirect objective measures	Laboratory experiments	e.g. Eye movements, reaction times
Direct subjective measures	Ratings of compliance	e.g. Ratings of compliance levels in given situations
	'Policy capturing' (can also be indirect)	e.g. Single ratings of complicance levels for multidimensional stimuli
Indirect subjective measures		e.g. Ratings of variables such as colour, signal word etc

colour. We intend making no judgements about which types of studies are most useful in warning compliance research. The reason for this is that there are numerous costs and benefits associated with each of the methodologies, suggesting that each can contribute to different aspects of the whole compliance issue. It is true that direct objective studies of warning compliance can immediately inform us of the effects of particular independent variables, but there are many data-gathering advantages to subjective studies which cannot, for practical reasons, be achieved with objective laboratory studies. An alternative taxonomy is provided by Ayres et al. (1992), and Lehto and Papstavrou (1993) provide a detailed summary of many of the recent studies in warning compliance, together with a classification of the type of study being reported.

In a number of studies compliance scores have been measured, and in some of these baseline (no warning) data are provided which allow the calculation of effectiveness scores. In the following sections some of these studies are reviewed, and compliance and effectiveness scores are given where appropriate.

Many studies on warning compliance have actually set up contrived situations where warnings are presented in varying formats and the level of compliance observed. Of course participants are always protected from real risk so either the task is one in which personal risk is relatively low, such as in the filing cabinet task of Frantz and Rhoades (1993), or the participants are not actually exposed to the risk, such as in the fake chemistry task used by Wogalter and his colleagues (Wogalter et al., 1987; Wogalter et al., 1989; Wogalter et al., 1993) where participants believed they were mixing chemicals were in fact just mixing coloured powders. In other instances, participants are actually stopped before they carry out a potentially hazardous task, such as that of using a circular saw (Otsubo, 1988) because the situation has been contrived to reveal the presence or absence of compliance before the actual task begins. In other cases, field studies have been carried out (for example, Wogalter et al., 1987; Dingus et al., 1993) where warning situations are manipulated, and compliance or noncompliance is noted. In some cases a hazard is contrived such as a broken door, or a faulty lift (Wogalter *et al.*, 1989), in other cases warning design is manipulated in situations where the behaviour to be observed is part of the normal procedure (such as Dingus *et al.*'s (1993) study of eye protection in squash) and where compliance may not normally occur.

Objective studies of warning compliance are valuable because they give us a direct measure of the effectiveness of a warning and thus allow us both to test hypotheses and to make comparisons between variables. They also allow us to observe or measure how likely it is that compliant behaviour will occur even when a warning is absent. We can therefore gain direct information both about the added value that the warning brings to the situation (the effectiveness score), and the baseline level of compliance behaviour (the compliance score), the bases for the two scores promoted in Chapter 1. From these measures we can gain empirical information relevant to decisions about warning use and warning calibration. Where a variety of variables are compared there will still be questions concerning the validity of cross-variable comparisons, and we shall turn to these problems in a later section, but objective studies certainly have the advantage of leaving fewer steps between the independent variables of concern and the ultimately important dependent variable of compliance behaviour.

A major advantage of objective studies is thus that they get straight to the issues at hand and remove the need to extrapolate from subjective findings to predicted levels of compliance. For example, Dejoy (1989) points out that several studies show there is a difference between simply noticing a warning and actually reading it, and there is also a difference between reading a warning and actually complying with it. Table 2.5, taken from Dejoy's review, contains the figures for noticing, reading, and complying from three of the studies in which both objective and subjective measures were taken. To an extent this has also been shown more recently in Frantz and Rhoades' filing cabinet study (1993), which shows that smaller percentages comply with than actually read the warning (interestingly in some cases more people complied with the warning than read it, showing that participants must have had previous knowledge about the dangers of filing cabinets). Ostubo's study is particularly informative because it also shows that the 'drop-out' is higher for equipment perceived as being less dangerous. For a dangerous piece of

Table 2.5 Percentages of people who noticed, read and complied with warnings in three studies (from Dejoy, 1989)

Study	Noticed	Read	Complied
	(%)	(%)	(%)
Friedmann (1988)	88	46	27
Otsubo (1988)	64.3	38.8	25.5
Strawbridge (1986)	91	77	37

equipment (a circular saw) about one-third of the participants were lost at each of the three stages. Of those who saw the warning, two-thirds read it; of those who read the warning, about two-thirds complied with it. For a jigsaw, a less dangerous piece of equipment, about half of the participants dropped out at each stage.

Despite their obvious advantages and empirical credibility there are some possible problems associated with objective studies, including the credibility of the task. Subjects have to be deceived into believing that they are at some risk when they are not. A task must be created which is both believable (by not being too risky or dangerous, otherwise the subject would probably assume that they are indeed being tricked) and yet also severe enough to induce compliance. Producing such a task presents the experimenters with some problems, but many of the objective studies appear to have succeeded in being plausible. In subjective studies, of course, the level of risk does not matter because participants are simply asked about these risks and hazards. Another limitation of objective studies is that only a small number of variables can be explored in one experiment, which makes cross-variable comparisons more complex than might be the case for subjective studies in which large numbers of variables can be considered. A third concern with objective studies is that apparently minor details of the design of the experiment might dramatically affect observed compliance rates. For example, if the subject is actually stopped before indulging in risky behaviour the experimenter cannot observe any compliance behaviour which may occur once the task has begun. In some tasks, for example, compliance may only occur during the task itself, and compliance might manifest itself in more cautious behaviour rather than in the actual use of, say, protective equipment. On the other hand, subjective studies are limited by all those factors usually limiting the application of subjective studies, so notwithstanding the limitations of objective studies we are likely to get somewhat closer to the truth in these than we are in many of the subjective studies.

In some of the earliest studies of warning compliance (Wogalter et al., 1987) participants were required to carry out a fake chemical mixing task. The variables under consideration were the format and location of a warning requesting participants to wear rubber gloves and a mask. A control condition was included, making it possible to calculate both compliance and effectiveness scores, as shown in Table 2.6. The compliance score corresponds to the proportion of participants complying with the warning as observed, but the effectiveness score considers the fact that some compliance was observed even when the warning was not present.

Thus the added value of the warning in this experiment was high. In a later set of experiments (Wogalter et al., 1989) no control conditions were used (although we can assume that baseline performance would have been much the same as in the earlier experiment, as the same task was used) and so only compliance scores can be calculated. Using the same chemical mixing task the experimenters showed the strong effects of cost and social influence on compli-

Table 2.6 Compliance and effectiveness scores partially recalculated from Wogalter et al., 1987)

	Warning at start	N at start	Warning at end	Warning at end	Control
Proportion compliance	0.9	0.7	0.5	0.5	0.1
Compliance score	90%	70%	50%	50%	10%
Effectiveness score	80%	60%	40%	40%	n/a

ance. Over three experiments the cost of compliance was manipulated by placing the protective gloves and mask further away from the participants, as well as by getting a confederate either to comply or not to comply with the warning. In the first experiment, which looked only at cost, compliance scores were 73 per cent in the low cost condition but only 17 per cent in the high cost condition. In the second experiment (low cost only) confederates influenced compliance rates to the extent that when a confederate complied, so did the subject at the rate of 100 per cent, but noncompliance on the part of the confederate reduced the compliance score to 33 per cent. In a high cost condition (the third experiment), confederate compliance produced a subject compliance score of 70 per cent, but confederate noncompliance produced subject compliance scores of 0 per cent. These results show not only the strong effects of both cost and social influence, but also suggest that the effects of social influence are greater than the effects of cost. Although true in this particular experiment, the question of calibration and equality between variables is an issue which needs to be clarified and will be considered later in the chapter.

More recent studies using the chemical mixing task (Wogalter and Young, 1991), which again did not include a control group thus making it possible only to calculate compliance scores, looked at the effect of print-only warnings, voice-only warnings and voice-and-print warnings. This experiment revealed compliance scores of 74 per cent in the print-and-voice condition, 59 per cent in the voice-only condition and 41 per cent in the print-only condition. Only the first and the last conditions were significantly different. A second experiment produced a much lower compliance score for the print-only condition, a rate which was similar to that found in earlier studies (Wogalter et al., 1989).

A more recent study still has looked at a more complex set of variables (Wogalter et al., 1993), again using the chemical mixing paradigm. Table 2.7 shows the conditions which were tested in this experiment. Hitherto virtually untested variables in behavioural studies such as visual clutter, strobe, and posted sign were included. Visual clutter was the presence or absence of clutter from the table on which a warning was placed. In the 'clutter' condition extraneous tools and electronic equipment were scattered around, whereas in the 'no clutter' condition only the warning plus the laboratory materials were

Table 2.7 Mean proportion compliance as a function of warning condition (from Wogalter et al., 1993)

Condition number	Condition description	Proportion compliance
(1)	Control – No warning – No clutter	0.111
(2)	Control – No warning – Clutter	0.000
(3)	Posted sign – No clutter	0.278
(4)	Posted sign – Clutter	0.111
(5)	Posted sign – Pictorials – No clutter	0.444
(6)	Posted sign – Pictorials – Clutter	0.167
(7)	Voice warning only – Clutter	0.611
(8)	Posted sign – Voice warning – Clutter	0.667
(9)	Posted sign – Voice warning – Pictorials – Clutter	0.722
(10)	Posted sign – Strobe – Clutter	0.222
(11)	Posted sign – Pictorials – Strobe – Clutter	0.278
(12)	Posted sign – Voice warning – Pictorials – Strobe – Clutter	0.833

Note. Control conditions 1 and 2 each had 9 participants.
All other conditions had 18 participants.

present. In the 'strobe' condition, a light was attached to a warning sign which flashed on and off eight times per second. Both the strobe light and a voice warning were activated when an infrared beam was broken by participants passing from the doorway to the laboratory table.

The proportion of compliance, equivalent to the compliance score, can also be seen in Table 2.7. This study used two control conditions so two different baselines can be established. Furthermore, the conditions are additive so it is possible to work out the 'added value' and effectiveness scores for specific variables. Table 2.8 shows the effectiveness scores (shown in proportions) for the variables tested, as well as the cumulative effectiveness scores of some combinations of the variables, calculated by us from the results obtained.

Significant and nonsignificant effects are discussed at length in Wogalter *et al.*'s paper. Because of the design of the experiment, it is possible to see the 'added value' of both particular variables and of sets of variables. We can see that the effectiveness of the posted sign is 0.111, or 11 per cent, the effectiveness score of the pictorial is 0.055, or 5.5 per cent, the effect of the strobe is 0.111 or 11 per cent, the effect of clutter is 0.111 or 11 per cent, and the effect of the voice message ranges from 55 to 61 per cent. Thus the presence of a voice message has a much greater effect than any of the other variables, for which the effects are much the same (and fairly low). There are a number of other features of these results which are worthy of discussion. For example, the effects for individual variables are remarkably consistent and appear to be independent of the other variables with which they appear. Consider the effect of the voice warning. When the voice message is presented together with the posted sign and clutter, the effectiveness score is 0.556; when presented with

Table 2.8 Effectiveness scores of individual and groups of variables, recalculated from Wogalter et al., 1993

Condition number	Description	Effectiveness score
1–2	Effect of clutter (no warning)	0.111
4–2	Effect of posted sign (clutter)	0.111
6–4	Effect of pictorial (posted sign + clutter)	0.056
6–2	Effect of pictorial + posted sign (clutter)	0.167
7–2	Effect of voice (clutter)	0.611
8–7	Effect of posted sign (voice + clutter)	0.056
8–4	Effect of voice (posted sign + clutter)	0.556
8–2	Effect of voice + sign (clutter)	0.667
9–8	Effect of pictorial (voice + sign + clutter)	0.055
9–6	Effect of voice (sign + pictorial + clutter)	0.555
9–7	Effect of pictorial + sign (voice + clutter)	0.111
10–4	Effect of strobe (clutter)	0.111
11–6	Effect of strobe (pictorial + sign + clutter)	0.111
11–10	Effect of pictorial (strobe + sign + clutter)	0.056
12–11	Effect of voice message (sign + pictorial + strobe)	0.555
12–9	Effect of strobe (voice + sign + pictorial)	0.111
12–10	Effect of voice + pictorial (sign + strobe + clutter)	0.611

the sign, plus the pictorial, plus the clutter, the effectiveness score is the same. The effectiveness score is also the same for the voice warning when it is presented with the sign, the pictorial and the strobe. The only anomaly – and it is a small one – is a slightly higher score for the voice when it is presented with clutter alone, where the value is 0.611. But the effects are very nearly identical. The same pattern is shown for some of the other variables. Take, for example, the effect for the strobe. When it is presented in clutter, the effectiveness score of the strobe alone is 0.111. This is the same when it is presented along with the voice warning, the sign and the pictorial, as well as when it is presented with the pictorial, the sign and the clutter. When one teases out specific 'added value' scores one can see that the effects of individual variables are remarkably consistent. Another feature of the data which only really shows itself when effectiveness scores are looked at is the additivity of the effects of variables. For example, the individual effectiveness scores of the posted sign and the pictorial are 0.056 and 0.111 respectively. When they are presented together, their combined effectiveness score is 0.167, which is exactly what their individual scores would predict. The same occurs for the effects of the voice and the sign. Individually, the effectiveness scores of these variables are 0.555 and 0.111. When they are presented together, their combined effectiveness score is 0.667, which again is the sum of the two individual scores. Some of the results are a little out of line, although not by very much, but generally they show that not only are effects of individual variables almost consistent regardless of the presence or absence of other variables, they also show that the effects are usually additive. The effect of two or three warning variables being present is almost equal to the sum of the individual effectiveness scores. In experiments such as this, we might begin to see how the effectiveness score might be put to good use. It can address directly questions such as consistency and additivity, two aspects of warnings-related variables which we need to know about if we are to begin to generalise from them.

The objective study by Otsubo (1988) discussed earlier also included a control condition in which no warning was present, which allows the calculation of effectiveness scores. In this experiment participants were led to believe that they would be required to carry out a carpentry task using either a circular saw or a jigsaw, the former is the more dangerous product. The main focus of interest was the format of the warning itself, which was either words only, a pictograph only, both pictograph and words, or a control condition where no warning was given. The measure of compliance was whether or not protective gloves were worn. In the case of the control group not one subject used the gloves, so the baseline was 0 per cent (in this case, therefore, the compliance and the effectiveness scores are the same). Table 2.9 shows the scores for actual compliance for the three experimental conditions.

Like Wogalter et al.'s 1993 study described above, the effect of pictographs appears to be quite low. Unlike that study, the effects of pairs of warnings do not appear to be cumulative, although there does appear to be some added effect for the pictograph plus words for the more dangerous piece of equipment, the circular saw. The most striking difference here is between the levels of compliance produced between the two pieces of equipment, being considerably higher for the circular saw. Contrasting with the effects of the design variables, the perceived danger of the product appears to have a much greater influence on objective warning compliance. From the questionnaire study carried out as part of this experiment it was found that familiarity with the product also affected warning compliance, with those participants who were less familiar with the equipment, and tools generally, exhibiting higher levels of compliance.

Among objective warning studies there are not very many that have included a control condition where no warning was present at all, making it

Table 2.9 Percentage of people who complied in each of the conditions in Otsubo's (1988) experiment

	Words only	Pictograph only	W + P
	(%)	(%)	(%)
Jigsaw	12.5	12.5	12.5
Circular saw	43.8	22.2	50

impossible to calculate an effectiveness score for particular conditions. In some cases where a control condition was used the complete lack of compliance in the control condition means that the compliance score is the same as the effectiveness score. There are many reasons why control conditions are not included (such as safety, for example, and the demands that might be put on the experimenter from external sources, such as in the Frantz and Rhoades (1993) study with filing cabinets) but their absence means that we should always be careful in drawing conclusions about the effectiveness of particular experimental manipulations *per se*. When no control group is used we can safely assume that there is only one (hypothetical) baseline for a particular experiment, but we cannot assume that this baseline would be the same for different experiments which might otherwise be comparable. For example, we know from those studies which have incorporated a baseline that familiarity with a product quite dramatically affects our willingness to comply (Otsubo, 1988). If we then were to compare two studies looking at similar variables, and found that the results differed, this could be because one study used participants experienced with a product and the other used inexperienced participants (such as psychology undergraduates). Similarly, it is also clear that the perceived hazardousness of the product has a large influence on warning compliance (for example, Wogalter *et al.*, 1987; Otsubo, 1988). Comparisons across experiments using products of differing hazardousness would therefore almost certainly be confounded by unknown differences in baseline compliance rates. Compliance scores could well be higher where the more hazardous products were used, but baseline differences could mean that the effectiveness scores might be lower.

Notwithstanding this baseline issue, objective studies make a valuable contribution to our knowledge of the effects, and relative effects, of variables thought to influence warning compliance. For the most part objective studies have focused on variables related to the design of the warning. For example, Strawbridge (1986) looked at the relative effects of warning position (with the warning at the top, in the middle, or at the bottom of the instructions and directions on an adhesive bottle), the presence or absence of highlighting of the warning message, and the placement of the 'critical' warning information (if the information about the cause, nature and prevention of the risk was either the first or the second sentence within a two-sentence warning section). There was a control condition of sorts, where the warning information was placed within the directions paragraph without being labelled as a warning, but there was no condition where information relating to the hazard was omitted. The actual behaviour required was that of shaking a bottle of adhesive to prevent skin irritation because of the acid allegedly contained in the bottle. In addition to the finding that the difference between the experimental and the control groups in recall of the information was negligible (which probably has a lot to do with the nature of the control condition, in which all the information was still given), the only factor to affect compliance behaviour significantly was that of the placement of the warning information. Compliance was greater (47 per cent) when the sentences specifying behaviour ('DANGER: Contains Acid. To avoid severe burns, shake well before opening') occurred before a sentence stating that the product was non-harmful and non-flammable than when they occurred after (27 per cent). It seems that when the critical information occurred second it was read slightly less, but more particularly it did not translate into behaviour so well.

In another more recent study Wogalter and Young (1994) looked at three alternative product labels, including the one usually used on their chosen product, which again was glue. Figure 2.4 shows the placement of the warnings in the three conditions. In the case of the 'Tag' and the 'Wings' condition the letters of the warning itself were larger, because of the increased space available for the warning. It is reasonably well established in the literature that font size itself affects factors such as perceived readability and perceived urgency (for example, Young *et al.*, 1992; Silver and Braun, 1993; Adams and Edworthy, 1995), so this may have had some impact on the results. The compliance scores for the three conditions were 13 per cent for the control condition (the manufacturer's usual label), 36 per cent in the 'Wings' condition and 80 per cent in the 'Tag' condition. Hence the type of warning had a very large impact on the degree of compliance obtained. Wogalter and Young attribute this finding to the added noticeability of the tag as compared with the other two conditions. It was visible from above, which was not true of the other conditions, and in addition it was visible throughout the task itself. This hypothesis was borne out by the results of the post-test questions about noticing and reading the warning, which mirrored the actual compliance

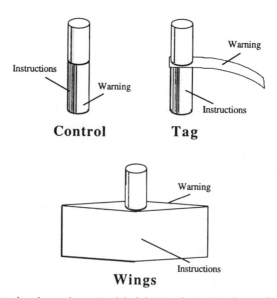

Figure 2.4   A control and two alternative label designs (from Wogalter and Young, 1994).

scores. It is interesting to note that in this particular experiment the compliance and the notice/read/recall scores were very similar, and that the drop-out observed in other studies (Dejoy, 1989) did not occur.

Braun and Silver (1995), in a follow-up to their study of signal-word/colour trade offs discussed earlier, carried out an objective study of warning compliance which was designed so that compliance scores could again be calculated. Again, an adhesive was one of the products to be used, together with a swimming-pool water testing kit. The warnings were presented in green, red, or black and participants were always told to wear rubber gloves. Colour showed a significant effect, with the colour red producing a compliance score of nearly 74 per cent, and green and black producing the same, lower, compliance scores (57.1 per cent) for the adhesive. For the pool testing kit compliance rates were slightly lower, being 68.4 per cent (red), 38.1 (green) and 47.6 per cent (black). Thus these results demonstrate that the colour red produces higher levels of compliance, as the subjective studies discussed earlier would also suggest.

2.6.2.2   Field studies

There is another distinct group of objective studies where compliance scores can be calculated, and where control conditions are difficult to set up. These are field studies, in which warning situations are contrived in natural settings (where there is no real danger or risk), and compliance with the warnings is observed. Many different situations have been contrived, such as broken photocopiers, broken doors, broken lifts, telephones and water fountains (for example, Godfrey et al., 1985; Wogalter et al., 1987; Wogalter et al., 1989). For example, in early studies (Godfrey et al., 1985; Wogalter et al., 1987) experimenters simply placed a number of warnings in different places and observed behaviour. In the first study (Godfrey et al., 1985), a simple warning message was placed on a photocopier stating that the machine did not work and that another one should be used. Not surprisingly, most people did not use the photocopier when the warning was placed on it, although some of them still did. In a second study a telephone had a warning label placed on it telling people that it was broken and that they would lose money. During the experimental session no participants were seen using the telephone.

Other experiments in this series looked at a water fountain and an ostensibly broken door, where again high levels of compliance were observed. In the case of the broken door, the cost of compliance was manipulated by varying the length of the detour which had to be taken and it was found that compliance rates fell dramatically as the cost of compliance increased, a finding which has been replicated (for example, Dingus et al., 1993). In a later study (Wogalter et al. 1989) the experimenters looked at the extent to which social influence affects compliance rates. In a study in which confederates either complied with or deliberately ignored a warning placed on a lift which was purported to be broken, it was first established that 33 per cent of all participants

ignored the warning and used the lift anyway. When a confederate was present, it was found that all but two of the participants complied with the warning when the confederate complied (producing 89 per cent compliance scores), whereas when the confederate did not comply compliance rates were 28 per cent, which did not differ from the baseline.

In another experiment a field study was set up using a voice message in a slippery floor scenario (Wogalter and Young, 1991). One of the features of this experiment not present in others was that a 'no warning' condition was used where the cues to activity and danger were present (cones, bucket and mop – although the cones themselves could conceivably be construed as a warning) as well as three other warning set-ups. In one, a printed warning was provided. In another, a voice warning was provided by a continuously looped tape, and in the final condition a voice-plus-print warning was provided, which was a combination of conditions 2 and 3. A criterion was generated for counting observed behaviour as compliant (not entering the defined area within 1.5 square metres of the cones) or otherwise. The presence of a baseline, where everything but the actual warning was included, makes it possible to calculate both compliance and effectiveness scores, which are shown in Table 2.10.

Thus in this experiment the voice-plus-print condition was most effective. If we look only at the effectiveness scores the effect comes closer to being additive than if we look at the compliance score themselves. If the effect were totally additive we would expect the print-plus-voice condition to produce a 66 per cent effectiveness score; in reality, the score was 56 per cent. If we look at the compliance scores alone this partial additivity is harder to see.

As well as providing some useful data, this study suggests a methodology appropriate for getting closer to effectiveness scores. There are a number of baselines which can be established, the most simple of which is the degree to which people comply if the evidence of risk is present but the warning itself is absent. This type of design gives us a useful methodology for looking at the effects of cues in the environment or on the product itself which will lead to compliant behaviour, (Chapter 1). It also allows us to look directly at the 'added value' of warnings, and to address questions such as the additivity of warning variables, another issue which we have explored.

There is another methodological issue which also needs to be considered. If a piece of equipment has a warning saying that there is something wrong with it, but because it is not actually faulty (as is true in many of these studies) the

Table 2.10 Compliance and effectiveness scores partially recalculated from Wogalter and Young, 1991

	No warning	Print	Voice	Print + voice
	(%)	(%)	(%)	(%)
Compliance score	20	42	64	76
Effectiveness score	n/a	22	44	56

subject may observe cues which indicate that the equipment is actually functioning, and so may go on to use the equipment assuming that the warning
label is perhaps out of date and should have been removed. This kind of
behaviour may inflate noncompliance levels over those which would be
observed when the equipment actually is faulty. So in studies of this kind it
may be particularly important to distinguish between compliance and effectiveness scores because of the possible causes of the observed baseline scores.

Other baselines, such as those taken in earlier studies, show the extent to
which people use a piece of equipment normally. For example, in situations
where the observer might be unsure if a person is going to use a piece of
equipment (such as a telephone) then baselines can be established both before
and after periods of observation, as these studies did. Establishing this type of
baseline is also important when looking at compliance scores.

Field studies are also particularly useful because they allow large numbers
of participants to be observed over a short period of time. This is useful in
warnings research because it is often the case that the experimenter or observer simply wishes to know if compliance takes place. Setting up an experiment
to ascertain this can sometimes be overly time-consuming and cumbersome.
As models of warning compliance develop it should be possible to make predictions about compliance levels given particular scenarios, and it is in the
testing of such predictions that field studies should come into their own. The
establishment of models of compliance behaviour is still a fairly long way off,
but we believe that field studies would be useful in such predictive models.

2.6.2.3 Subjective studies

There are also a large number of subjective studies of warning compliance,
covering a wide range of topics. In some cases these subjective studies look at
perceived, rather than actual, compliance rates (for example, Wogalter et al.,
1987; Wogalter et al., 1991; Silver et al., 1991; Polzella et al., 1992; Laughery
et al., 1993a) usually along with other subjective variables known to influence
eventual compliance, such as perceived hazardousness of the products or
labels under consideration. Other subjective studies have not looked explicitly
at perceived compliance, but at those variables thought to influence compliance. These have usually been studies of design variables such as font size and
perceived readability. (For example, Young, 1991; Young et al., 1992; Malouff
et al., 1992; Laughery et al., 1993b).

Subjective studies, although they cannot address the issue of compliance
directly as can objective studies, have many advantages. They allow the comparison of many different variables within a single experiment (although the
question of the validity of cross-variable comparison still remains), they allow
data to be collected more quickly, and they of course allow hazards and scenarios to be presented to participants that would not be allowed for safety
reasons in objective studies. Many techniques have been used in subjective
studies, only some of which will be reviewed here. What follows is not an

exhaustive literature review as studies are only included if they directly address one or more of the issues considered in this chapter.

A line of research which has not been pursued as much as it might have been is that originally presented by Wogalter *et al.* (1987) where participants were asked to rate 17 different warnings, covering a range of hazards differing in their degree of severity. The main focus of interest in this study was the relative effectiveness of the four prescribed parts of a warning, which are the signal word ('Warning', 'Note' and so on), the hazard statement, which tells of the actual hazard (such as overhead cables and poisonous liquids), the consequences of noncompliance (for example, electrocution and skin irritation) and instructions (telling the user or observer what or what not to do in order to avoid the hazard). The results of the study are interesting in themselves, but the methodology could be very usefully employed in the determination of effectiveness scores.

Wogalter *et al.* (1987) presented participants with the 17 warnings both in the full four-statement version and in all permutations of the possible three-statement warnings (such as one containing only the hazard statement, the consequences and the instructions, omitting the signal word). Subjects gave subjective measures of the perceived effectiveness of each of the warnings, as well as a redundancy measure for each of the separate statements, which complemented the perceived effectiveness measure. The results showed that warnings in which all four statements were present were perceived as being more effective than any of the three-statement versions. Furthermore, they found that greater reductions in perceived compliance scores were obtained when either the hazard statement or the instructions were removed. This suggests that the parts of the warning which relate directly to behaviour, i.e. the parts which tell the user what to do, are rated as being the most effective, whereas the parts which give background supporting information (such as the consequences) and the merely alerting or iconic parts (the signal word) are rated as being less effective in the overall warning. There is other evidence, to be reviewed later in this section, suggesting that the degree of explicitness of a warning affects perceived compliance rates (for example, Laughery *et al.*, 1993a) which again concerns the informational part of the warning.

A relevant methodological point in this experiment is that participants always knew the stimulus they were looking at was a warning, and that no stimuli were presented which were not warnings. In these circumstances, where participants already know that they are looking at warnings, we may expect results to show that signal words are redundant. It would be interesting to find out if signal words are seen as redundant in studies where participants do not necessarily know that they are looking at warnings. A more substantial methodological point is that this study allows us to calculate compliance scores but gives no possibility of measuring effectiveness scores for individual warnings. Wogalter *et al.* (1987) found an overall effect for hazard, in that those situations which warned of greater or more severe risk were perceived as being more effective warnings overall. This tells us that the compliance scores in these situations are higher, as much of the warnings literature might predict. It does not tell us what the warning itself adds to the baseline level of compliance that would be produced anyway. For example, in situations where even without a warning the risk is well known or obvious and high, compliance scores are also likely to be high and an explicit warning may add little to the situation. On the other hand, a situation where the risk is not well known or obvious may benefit greatly from an explicit warning.

Subjective methods such as those discussed enable direct questions about the effectiveness of warning labels and signs to be asked. To our knowledge, such experiments have not been carried out. For example, participants could be shown a variety of products, scenarios, or even photographs of situations in which risk is present. The control condition would be the stimulus without any warning sign or label, and the experimental situation would be the one where the warning is also present. Direct measures of effectiveness could then be calculated by observing the differences in the perceived risk (or some other subjective measure of the participants' willingness to indulge in compliant behaviour) with and without the warning sign or label being present. Such a technique would reveal a number of useful pieces of information. First, it would allow us to distinguish between the situation itself with all its inherent cues, and the warning which represents it. Secondly, it would allow us to see directly if different levels of hazard (which can be judged objectively by experts within the particular domains represented by the warnings and situations) produce different baseline levels of performance, something which many subjective studies already tell us, but not quite in these explicit terms. Finally, it would allow us to examine the relative effects of different variables on the effectiveness of warnings because we would be able to subtract the effect of the warning from baseline performance. In our view, this kind of methodology would produce interesting and important results both for research and practice. Data from studies such as this could directly address the 'but for' issue commonly encountered in legal cases. Subjective data on the 'before' and 'after' effects of warning provision would be relevant to decisions about similar cases encountered in everyday life.

One technique sometimes used in subjective studies is that of 'policy-capturing', (Zedeck and Kafry, 1977), a technique relying on regression techniques which is often used in market research. This technique allows the relative contributions of what are normally a large number of variables to be assessed; the initial selection of levels of those variables is of course another issue, which is discussed at another point in the book.

Schwartz et al. (1983) used the policy-capturing technique not to address the question of warning compliance directly but to look at buying intentions in which hazard information figured. Pilot studies revealed the factors most likely to influence buying decisions relating to three products – spray oven cleaner, liquid bleach, and drain cleaner. These factors were isolated as cost and effectiveness, and hazard was included because it was the main focus of the experiment. Subjects were presented with scenarios in which the effectiveness, cost and hazard were quantified, and participants were asked to rate the degree to which they were likely to buy the product. Data analysis, using regression techniques, showed that both hazard and effectiveness contributed to participants' perceived purchase intentions, with greater hazards decreasing the likelihood that people would buy the products. Regression techniques also allow trade offs to be seen, and in this study it was found that participants were willing to pay more for a product which had a lower hazard probability. In this particular instance it could be calculated that participants were willing to pay $1 more for a product which was three units safer on a hazard scale (falling from a hazard probability of 1 in 300 to one of 1 in 600).

Another study also largely using this technique was carried out by Ursic (1984), who manipulated three design variables and asked participants to rate the effectiveness and safety of three hypothetical brands of insect spray and hair dryers. The variables manipulated were lettering case (upper/lower), the use of a pictogram (whether it was present or absent), and the signal word which was 'Danger', 'Caution' or was left out altogether. After observing the products, participants were asked about their perceptions of the product one hour later. One of the results was that products were perceived as being significantly safer when presented with, rather than without, a warning label. This shows us the extent to which warnings labels have become integral parts of products and situations themselves, a conclusion which is perhaps even more true now than when this study was conducted. However, the design manipulations did not affect the perceived effectiveness or safety of the product.

There is no reason why the policy-capturing technique could not be applied specifically to the question of warning compliance, although there appear to be no studies which do this directly. For example, the relative contributions of a whole range of different variables to a respondent's willingness to comply with a warning could be ascertained by asking him or her, on the presentation of composite stimuli such as those used by Schwartz et al. (1983) and Ursic (1984) to rate the degree to which they (or the population at large) would comply. Regression techniques would then show which of the variables have strong effects, and which have smaller effects. Such techniques, if developed, could be of great use in determining the important variables in warning compliance because they would allow examination of the relative contribution of a large number of variables to a single decision, that decision being an estimate of the degree to which the subject would comply with the warning, or some similar measure.

There is another sub-group of subjective studies which asks participants to rate one, or several, dimensions thought to be related to warning compliance (see for example, Wogalter et al., 1986; Laughery et al., 1991; Silver et al., 1991; Wogalter et al., 1991; Polzella et al., 1992; Laughery et al., 1993a). In these cases the stimuli are not usually composite; it is the participants' responses, rather than the stimulus materials, which are multidimensional. For example, Silver et al. (1991) took 26 household pest-control products and asked participants to rate them on a 0–8 scale on a number of measures including how hazardous they believed the product to be, how familiar they were with the product, how understandable the warning was, how willing they would be to read the warning, and how difficult it would be to use the product. Of particular interest in this study was the relationship between participants' willingness to read the warning label and other factors. It was found that willingness to read the warning label correlated with product hazardousness, the understandability of the warning, the attractiveness of the warning and other hazard-related variables. They also found that participants' willingness to read warning labels was correlated with some objective readability measures such as the number of statements in the warnings and the Flesch (1948) grade-level index. Regression techniques showed that the hazardousness of the product accounted for nearly half of the variance associated with people's willingness to read a warning. Warning understandability and attractiveness accounted for significant amounts of variance, but others, including familiarity, did not. A regression model containing all three predictors accounted for 96 per cent of the variance. If we assume that there is a relationship between a person's initial action of reading a warning and their eventual compliance (studies such as those of Otsubo (1988) suggest this is the case) then we can extrapolate these findings by arguing that hazardousness, together with understandability and attractiveness of the warning, are important factors in warning compliance, although of course it is not necessarily the case that we have distinguished compliance behaviour from warning effectiveness. Earlier studies (Wogalter et al., 1986 and Wogalter et al., 1991) also indicate that people's willingness to read warnings is closely related to the perceived hazardousness of the product under consideration. This suggests that under many circumstances it is the compliance score we should be interested in because the product and the warning label are inextricably joined together, and under some circumstances it makes little sense to try to separate the two.

Another variable which has shown itself to be important in determining the degree to which a person judges him- or herself to be likely to comply with warnings, or to otherwise exhibit cautious behaviour, is that of the explicitness of the information contained within the warning label (Laughery et al., 1991; Laughery et al., 1993a). This is a factor of some importance because it focuses directly on the informational part of a warning rather than on its iconic aspects. What these studies tend to show is that the more explicit the warning label the more likely are participants to exhibit careful, compliant behaviour. Studies of this type tend to compare different types of warning labels. The nature of the studies does not readily allow the inclusion of a condition with no warning label, so that from these studies it is not possible to calculate effectiveness scores. These studies typically focus on purchasing behaviour as it is often a worry of product manufacturers that the effect of providing explicit hazard information (which may promote more safe behaviour with respect to the product) may put people off buying that product, a research finding reported by Silver et al. (1991).

The whole issue of the effect of explicitness in a warning has been explored in a recent research study (Laughery *et al.*, 1993a) which we will look at in some depth. In the first of the four studies reported in this paper the experimenters tried to distinguish between explicitness and severity to discover to what extent these two features are orthogonal to one another. Subjects rated a large number of warnings on explicitness (defined as knowing exactly or specifically what will happen to you) and severity (defined as how bad or injurious the consequences are). Four categories of warning type were generated, as shown in Table 2.11. The four 'best' warnings (one for each of the four products) in each of the four groups – nonexplicit/nonsevere, explicit/nonsevere, nonexplicit/severe and explicit/severe – were then tested by having participants complete a questionnaire asking them about eight features of the products. These features included hazard level, likelihood of injury, severity of injury, intention to act cautiously, product familiarity, catastrophic level of risk, likelihood of reading a warning, and control over potential hazards.

The results showed that the two dimensions, severity and explicitness, are not clearly orthogonal, although a fair degree of distinction between the two is possible. Figure 2.5 shows that to some extent the two are positively related. One of the most interesting findings from this experiment was that nonexplicit warnings reduced participants' intention to act cautiously in relation to the product. Warnings which were both nonexplicit and nonsevere were generally found to score slightly lower on the severity dimension than those which were explicit but nonsevere. One may have expected the reverse – that explicit nonsevere warnings would be perceived as being less severe than a comparably nonsevere warning for which the hazard is not clearly stated. For example, if one were using a product with a warning that was not very clear, but which implied that the product was hazardous in some way, then one may predict that one would be more careful with it than with a comparable product where the low level hazard was stated quite explicitly. However, Laughery *et al.*'s

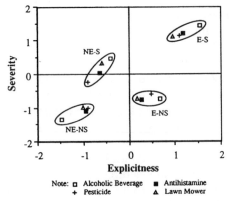

Figure 2.5 Plot of 16 warnings by explicitness and severity ratings (from Laughery *et al.*, 1993a).

Table 2.11 Warnings used in Laughery et al.'s, 1993a experiment. Severity is graded as severe or nonsevere (S or NS) and explicitness is graded as explicit or nonexplicit (E or NE)

Product	Warning
Alcohol beverage	
NE–NS	If you drink alcohol regularly, it is not good for you. (E = −1.48; S = −1.33)
NE–S	Mixing alcohol and medicine can be life threatening. (E = −0.41; S = 0.46)
E–NS	If you drink alcohol, it begins to reach your brain within two minutes after drinking it. (E = 0.67; S = −0.73)
E–S	If you drink while you are pregnant, your child may be born With Fetal Alcohol Syndrome and need institutionalization. (E = 1.55; S = 1.45)
Antihistamine	
NE–NS	Some may be allergic to this drug, which may result in discomfort. (E = −0.96; S = 1.08)
NE–S	Using this drug while drinking or taking other drugs may result in severe effects. (E = −0.64; S = 0.05)
E–NS	This drug may cause drowsiness, which may result in minor nausea and dizziness. (E = 0.27; S = −0.75)
E–S	Exceeding the recommended dosage may result in brain damage, prolonged coma or death. (E = 1.17; S = 1.20)
Chemical/pesticide	
NE–NS	Ingestion of this product may result in discomfort. (E = −0.90; S = −1.01)
NE–S	Exposure to the skin may result in extremely serious problems. (E = −0.90; S = −0.23)
E–NS	Exposure to the skin may result in skin irritation, dry skin and/or minor skin discoloration. (E = 0.49; S = − 0.59)
E–S	Inhalation of this product may result in major lung burns, lung collapse and death. (E = 1.11; S = 1.13)
Lawn mower	
NE–NS	Touching the muffler or engine after use may result in discomfort. (E = −1.01; E = −0.97)
NE–S	Putting any part of your body near the moving blades may result in serious injury. (E = −0.59; S = 0.33)
E–NS	Breathing the fumes from the mower exhaust may result in minor dizziness and nausea. (E = 0.17; S = −0.73)
E–S	Allowing a match or cigarette to come near the mower may result in major fire or explosion, which can cause serious burns or death. (E = 0.95; S = 1.11)

(1993a,b) results do not show this. It is, however, at the severe end that the difference in severity ratings becomes really noticeable.

Differences in severity scores are much higher for explicit than for non-explicit warnings, even though the levels of actual severity are equal. Thus it seems that explicitness generally tends to increase the perceived severity of risk in a product, particularly so at the 'severe hazard' end. This effect was further demonstrated in the second experiment, where a variety of products were assessed along six dimensions including severity of risk, buying intentions and explicitness. The key factor here was that each product label (which ranged from a motor cycle helmet to drain cleaner) was presented in explicit and nonexplicit forms. Again, explicit warnings produced higher ratings of injury severity than the nonexplicit warnings.

In the third and fourth experiments in Laughery et al.'s (1993a) study purchasing intentions were looked at more explicitly, and there was detailed examination of other factors such as cost and quality, which might also affect purchasing intentions. The results do show to some extent that greater explic-itness in warnings adversely affects purchasing intentions, but the results are slightly ambiguous.

The finding that explicitness and perceived severity are correlated presents an interesting implication in terms of the calibration issue raised earlier in this chapter and Chapter 1. Explicit warnings convey a much higher degree of severity for consequences which actually are severe, and there is a smaller effect for nonsevere hazards. Perhaps the recommendation drawn out of these studies is to use nonexplicit warnings for nonsevere hazards (to reduce the perceived severity of the hazards) and to use explicit warnings for severe hazards. Indeed, the rule might be to increase the explicitness of the warning monotonically as the severity of the hazard increases.

2.7 Variables and the problem of generalisation

Studies which involve subjective measures, such as those concerned with deter-mining the most important variables in warning compliance, allow a large number of variables to figure in their design. When a large number of vari-ables is included in the one design the relative contribution of each variable to the overall judgement or judgements being made can then be ascertained. Our feeling about analysis is that these relative contributions are easier to see if regression models are used, but other approaches are of course possible. Analysis issues aside, when a range of warning-related variables is being exam-ined it could be because we are interested in the warning as part of the product or situation, in which case we would be concerned with compliance scores. On the other hand, we could be interested in variables concerned with features of the warning itself, separate from the product or situation, in which case we would be concerned with effectiveness scores. Some studies measure one and some studies measure the other, which sometimes clouds the issue as to the relative impact of variables on behaviour. For example, although it can be shown almost directly from objective studies that certain variables related to the warning design (such as providing a verbal warning) can have a large impact on behaviour, these studies do not tell us that the perceived level of hazard can have a large impact on our predisposition towards careful compliant behaviour in the first place, as demonstrated by subjective studies. The relative contributions of warning design variables and of hazard level variables are therefore hard to disentangle, and this problem is sometimes confounded by the lack of distinction between compliance and effectiveness scores.

On top of this, even if meaningful comparisons can be made across variables because we have disentangled compliance scores from effectiveness scores, we still have the problem that, because of the way they have been selected, cross-variable comparisons may not be as valid as we think they are on first sight. An argument for a consistent philosophy in the selection of variables was made in Chapter 1, which we will not go over again. The point we do wish to make here is that in some variables such selection is a relatively easy process, whereas in others it is a veritable minefield. Unfortunately, it may be the case that those problem areas are precisely where we should be looking to isolate those variables having a large impact on warning compliance. Two examples are given here, which in many ways use methodologies at two ends of the spectrum. The first, a study of the perceived urgency of warning label attributes (Adams and Edworthy, 1995) measures subjective judgements about clearly delineated physical variables. The second (Wogalter et al., 1989) measures objective levels of compliance in a social situation, where the variables are much harder to delineate. In both cases the experimenters have sought to make direct comparisons concerning the relative contributions of independent variables to changes in the dependent variables under investigation.

Adams and Edworthy looked at the relative contributions of font size of the word 'Warning', border width (the thickness of the border surrounding the word 'Warning'), the amount of white space between the signal word and the border, and the effect of colour (black vs. red) in judgements about the perceived urgency of written warning labels. The actual warning label used is shown in Figure 2.6, a version of a generic product label. The main purpose of the experiment was to explore the relative contributions of each of these variables to overall judgements about urgency. Numbers were assigned to subjects' estimations of perceived urgency, and a regression technique was used to quantify the relative strengths of these variables.

The generation and selection of variable levels was central to the aims of the experiment, and was done in the following way. Stimuli were presented to participants on cards simulating the natural size of labels of this sort, which might appear on dangerous liquids. As it is intended for use on a bottle, there is a maximum size to the complete label. There is limited amount of space available on the label so there is a limit to the maximum size in which the word 'Warning' can be printed in order that the rest of the information is

WARNING

Wear protective waterproof clothing, cotton overalls buttoned to the neck and wrist, washable hat, elbow length PVC gloves, impervious footwear and full face respirator with combined dust and gas cartridge. Avoid contact with eyes, skin and clothing and avoid inhaling vapour. Before eating, drinking or smoking wash hands and face thoroughly with soap and water. Avoid any contact at all with this substance.

Mixing
Use measuring beaker provided. Add the required quantity to water and mix thoroughly. Do not mix more spray than can be used on the day of mixing.

Storage and disposal
Store in the closed, original container in a cool, dry place. Do not store in direct sunlight. Dispose of empty container by wrapping in paper then put in a plastic bag and place in garbage.

Figure 2.6   Warning label showing mid-values of type size, white space and border width (from Adams and Edworthy, 1995).

readable. There is also a lower limit, below which the word is not readable. Thus the whole practical or usable range for font size, using this particular experimental set-up, was from 8 to 32 points. In the first of these experiments, seven levels of font size were chosen, equally spaced between these two limits. In the case of border width, the smallest width which can be used is 1 point, and aesthetic judgements taken of border widths within the overall size constraint suggest that border widths greater than about 8 points are quite unaesthetic, and so would be very unlikely to be selected by designers, given the other size constraints. Thus the usable practical range for border width, under these circumstances, was from 1 to 8 points. The amount of white space between the signal word and the border was varied from 2 to 32 points, with 2 the smallest possible and 32 the largest possible within the overall size constraint of the whole label. Thus in all three of these variables the whole of the practical, or ergonomically feasible, range of values was covered. This means that comparisons across variables become meaningful, and designers can draw conclusions about the effects of these variables when designing warning labels. Warning signs and labels can be larger than those which were used but the proportions would be likely to stay much the same.

The results are shown in Figure 2.7. The clearest effects are for text size and border width, with white space having very little impact across the seven variable levels. From the graph we can infer that text size has a greater effect on perceived urgency than border width, and both have greater effects than the amount of white space. The regression coefficients for each of these three variables are shown in Table 2.12. Two regression coefficients are shown for each variable. The one based on absolute values shows a higher coefficient for border width than for font size. This means that, in absolute terms, a unit increase in border width has a greater impact on perceived urgency than a unit increase in font size. However, border width can only vary effectively from 1 to 8 points, while under the same circumstances font size can vary from 8 to 32 points, so although this coefficient is interesting, it does not tell designers what they really need to know. The abscissa of the graph (Figure 2.7) is labelled in independent variable units, rather than absolute units, for this same reason.

We argue that the really meaningful regression coefficient is the one which includes these nominal values because it treats the two ranges of text size and border width (and white space, which has little effect) as equal, which they are in a very practical way. Using the nominal values, text size has a greater

Table 2.12 Absolute and nominal regression equations for the effects of white space, text size and border width on perceived urgency (from Adams and Edworthy, 1995)

Variable	Absolute x values	Nominal x values	Variance accounted for (%)
Text size	$y = 6.3x + 20.6$	$y = 25.3x + 45.5$	99.2
White space	$y = 819x + 135.4$	$y = 4.1x + 133.0$	71.1
Border width	$y = 10.7x + 101.8$	$y = 12.4x + 100.3$	93.5

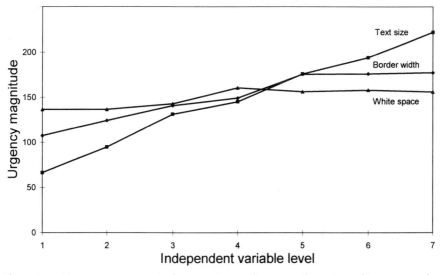

Figure 2.7 Mean urgency magnitude estimations as functions of text size, white space and border width (Adams and Edworthy, 1995).

regression coefficient than border width. This means that within the practical constraints of the design possibilities variation in text size will have a greater impact on perceived urgency than border width. The ergonomically feasible range for each variable is different, and these differences should be considered if meaningful comparisons are to be made across variables.

The variables examined in this last study were chosen because they are relatively easily quantified. There are other important variables, however, for which such quantification and comparison is much more difficult. Earlier in this chapter we discussed an experiment by Wogalter *et al.* (1989) in which objective levels of compliance with warnings in the chemical-mixing task were measured with respect to the effects of cost and social influence. The first experiment showed that when the cost of compliance was high there was very little compliance, but when the cost was low nearly all participants complied. In the second and third experiments both cost and social influence were varied. The relative effects of the two were directly compared by plotting the means found in the four conditions tested in this experiment. These were low cost/confederate compliance, low cost/confederate noncompliance, high cost/confederate compliance and high cost/confederate noncompliance. The means are shown in Figure 2.8. From this comparison the authors conclude that social influence appears to have a greater effect on compliance behaviour than cost – in this experiment the effect of social influence was twice that of cost. Although this conclusion follows from this particular study, there would be problems in generalising from this observation to situations where the details were different. We can begin to see how difficult it would be to predict the size of the effect a variable would have on compliance behaviour even though we could reasonably expect the effect to be a strong one. For example, the study

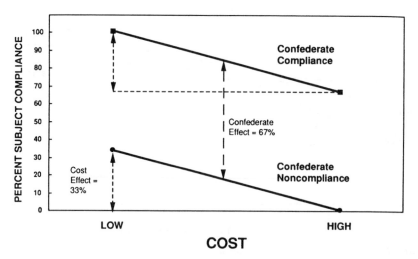

Figure 2.8   Relative effects of cost and confederate compliance (from Wogalter et al., 1987).

we have just discussed showed that the presence of one confederate produced a strong influence on the subjects' compliance behaviour, but we do not know what the added effect of two or more confederates would have been. Evidence from social psychology, however, (for example, Asch, 1955) suggests that the effect reaches asymptote after about three confederates. Thus with more confederates the effect of confederate compliance might become complicated.

Another variable examined in this same study was cost. There are a number of problems, two are considered here. First, cost was equated with the distance that the subject needed to travel to comply with the warning. There are many other potentially quantifiable variables that can be thought of as cost variables, all of which can be examined using utility and decision-making models as discussed in Chapter 1. The time needed and the difficulty encountered in complying (Dingus et al., 1993) are other possible cost variables. Thus cost can be operationalised using many different dimensions, but more study is needed on how cost can be measured in warnings research. Secondly, the calibration of the cost variable is of some interest. Dingus et al., (1993) have suggested that small increases in cost can, in their words, 'devastate compliance rates' which may not be as true in Wogalter et al.'s (1989) study. This implies to us that the Dingus et al. (1993) study covers a wider range of cost values than does the Wogalter et al. (1993) study, although quantifying this claim is impossible. It might also be true that the perceived hazard in Dingus et al.'s study is lower than that in Wogalter et al.'s (1993) study, which would again reduce compliance rates. Developing techniques whereby cost can be quantified in some more linear and comprehensive way is again something which might be explored in future research work.

Thus we may imagine some other study where cost has a greater effect on compliance, and where it might even outdo confederate compliance in its effects. Such comparisons across variables are useful, but we need to consider the generalisation problem, which in our view is tied in with the variable range argument discussed here and in Chapter 1. In the Adams and Edworthy (1995) study the problem is neat and contained and readily quantifiable; in the Wogalter et al. study, where the variables themselves might ultimately show themselves to be of greater significance in compliance, the problem is much more open-ended and complex. However, we believe that the variable range problem really needs to be tackled if we are to increase the extent to which we can generalise from specific warnings studies and to work towards generating predictive models of warning compliance.

CHAPTER THREE

Symbols

3.1 The language of symbols

As we have seen in Chapter 2, verbally-based visual warnings convey their message through their four components. These are a signal word such as 'Warning', usually presented in a way designed to gain attention, a hazard statement indicating the actual hazard, a statement indicating the consequences of noncompliance and some instructions telling the user or observer what to do, or what not to do, to avoid the hazard. One or more of these components can be replaced with a symbol, and it is to the use of such symbols that this chapter is addressed. In this chapter we will largely be concerned with understanding the informational content of symbols as replacements for the informational aspects of written warnings, but of course they can also fulfil iconic roles, which we will also discuss.

The history of written communication begins with the use of pictorial symbols to represent simple ideas, but the simple stand-alone symbols that were originally used soon gave way to abstract writing systems such as cuneiform and hieroglyphics (Senner, 1989). It was very early in this process that the pictorial representations and the way they were used became so stylised that, as with our modern alphabets and syllabaries, the communication system could only be used by those who had learned the meaning of the symbols and their system of use. Now, spurred on by the extent of our international contacts, we have returned to the situation where there is a need to communicate simple concepts such as roadway information, other matters of public information and of course product warnings, quickly and efficiently in ways not requiring extensive learning. We thus find ourselves returning to the use of simple graphic symbols where appropriate.

Our modern use of symbols in areas such as public information, industrial safety, and now also in the world of computer interfaces, follows on their great success in the area of highway signs. The first attempt at standardising highway signs was at a 1909 international conference on highway sign systems where four symbols (for curve, bump, intersection and railway crossing) were adopted, with no prescription for sign border shape or colour (Eliot, 1960). A Paris convention in 1926 established the triangular shape for danger signs and, in 1931, the League of Nations expanded the symbol series to 26 signs and applied some colour and shape standardisation – regulatory signs were to be circular, information signs rectangular, and red was to be used as the predominant colour for prohibitory signs (Eliot, 1960). Finally, the United Nations Protocol on Road Signs and Symbols of 1949, revised in 1953, formed the basis of the standardised road symbols now used almost universally in Europe and widely adopted elsewhere.

Many of the signs that were first used in the road context have been used without change in other areas. The exclamation point for warning, and the stop sign, are examples. Most of the research on the use of symbols has been with highway signs, but there has been some research on public information symbols, machinery symbols and also on symbols used to indicate hazards.

Modley (1966) and Dreyfuss (1972) have both pointed out that there are three categories of symbol: the image-related or representational, the concept related or abstract and the arbitrary. Examples of these are given in Figure 3.1. A pictorial symbol is a direct representation of an easily recognised object

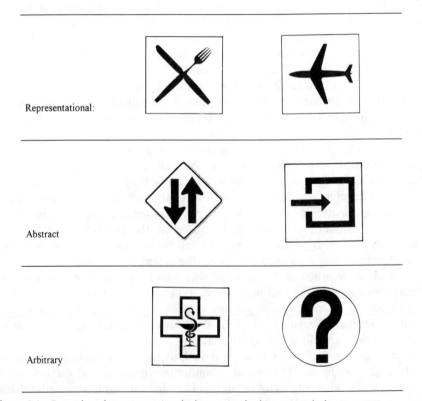

Figure 3.1 Examples of representational, abstract and arbitrary symbols.

or feature. Modley (1966) pointed out that representational symbols tend to change in time, so that, for example, a 'no horns' road symbol showing a klaxon horn with a slash through it would be meaningful to an older generation but probably not to current road users. Yet, where a suitable image does exist, that image can provide a powerful and unambiguous evocation of the object depicted. It does not, however, guarantee that the meaning of the *symbol in context* will be clear.

Concept-related or abstract symbols are those where some aspect of a concept lends itself to symbolisation in such a way as to evoke the concept readily. Thus a right-turning arrow suggests right turn. Indeed, the idea of placing a slash to indicate prohibition is itself a concept-related notion.

Arbitrary symbols are those where the meaning is arbitrarily assigned to the graphic. Thus an exclamation mark does not depict any aspect of the real world object, but has a meaning that has been arbitrarily assigned to it and that we have had to learn to understand without any assistance from representational elements within the graphic itself. Where such arbitrary symbols are used the graphic image should be simple and clearly distinct from any other symbol that might be used in a similar context.

3.2   Why use symbols?

Sometimes symbols are used without explicit consideration of the reasons why. It is likely that conscious consideration of the reasons for using symbols will result in a more intelligent application of symbols. The following presentation gives those reasons together with appropriate evidence.

3.2.1   Symbols can be recognised by those who do not read the vernacular

The most obvious reason for using symbols relates to increased international travel and communication which are facilitated if some basic concepts can be communicated in a way understandable by those from different language backgrounds. That is, if appropriately chosen, a symbol permits communication with someone who does not read the language of the country. For example, Cairney and Sless (1982) showed that symbols that were effective with native-born Australians were equally effective with recent arrivals from Vietnam. There is also a need to communicate warnings to those who may be illiterate in any language. Definitions of illiteracy vary widely, but particularly within industrial populations the extent of illiteracy is commonly underestimated.

3.2.2   For signs of the same size the symbol version can be recognised from a greater distance

In the realm of road signs there are many examples of symbol and verbal signs that are of equivalent meaning. Jacobs, *et al.* (1975) have shown that for 16 such pairs of signs the symbolic version is, on average, recognised at twice the distance that is required for recognition of the equivalent verbal sign of the same overall size. Their recognition criterion was the visibility distance required for 50 per cent accuracy, but they also noted that as the sign's distance decreased the symbol signs approached the 100 per cent asymptote more quickly than the worded version. Only when the worded sign involved highly distinctive features was it recognised at approximately the same distance as the symbol version. This applied to signs such as 'No entry' and 'Slippery when wet'. The authors suggest that the participants were so familiar with these verbal signs that secondary cues such as word layout provided sufficient information to identify the sign correctly. Kline *et al.* (1990) investigated the visibility of equivalent symbol and worded signs in the light of the poorer contrast sensitivity of older viewers. Their results showed that the visibility of symbol signs is better than verbal signs not only for older viewers but for viewers of all ages. This advantage was present during daylight viewing conditions but was greater under the reduced lighting of dusk viewing conditions.

3.2.3 Symbols can be recognised more quickly and more accurately than the equivalent worded sign

In one of the earliest pieces of research on symbol signs, Jander and Volk (1934) compared reaction times to both symbol signs and word signs that had a directional component. They had participants push a lever in the direction indicated by the symbol or sign. They found that reaction times were shortest to the symbol version, and that the difference, which was in the order of 200 m s, increased with practice.

Walker *et al.* (1965) investigated differences between American worded road signs and international symbol signs. Participants were given the chance to study the symbol and word signs and were then shown them in random order for 0.06 seconds each, with the request to write down what they saw, i.e. whether it was symbol or words, and the meaning of the sign. Results showed large differences in favour of more accurate identification of the symbol signs.

King (1971) used viewing times of from $\frac{1}{3}$ to $\frac{1}{18}$ seconds to assess the legibility of 18 symbol and 9 worded signs under brief viewing conditions. His outcome measure was the percentage of correct matches between each briefly-presented symbol or word stimulus and an answer chosen from an array of symbols or words. Overall, the percentage of correct matches was greater for symbols than for word signs, but more important was the finding that for word signs the percentage of correct matches decreased as the exposure time decreased, and that for symbol signs there was no such decrease. Furthermore, 65 per cent of the participants stated that they found the symbol signs easier to match than the word signs.

Dewar *et al.* (1976) found, in a series of studies, that when participants were asked to say what a sign meant the response was faster to worded signs, but they point out that the response requirement involved additional cognitive processing for the symbol signs. Thus for the verbal signs all that was needed was to read the sign, whereas for the symbol signs it was necessary to generate a response. However, under conditions of visual distraction the advantage of worded signs disappeared. Ells and Dewar (1979) used a different measurement method. This time they read out the verbal name of the sign, e.g. 'two-way traffic' and then showed a slide of a verbal or symbolic sign which did or did not correspond to the message. Participants were simply asked to say 'yes' or 'no' to indicate whether the displayed sign was the same or not. Under these conditions the response was faster to the symbol signs.

3.2.4 A symbol sign can withstand greater degradation and still be recognisable

King (1975) in a later experiment introduced delays of 5 and 10 seconds before responding was permitted. This occurred both with and without an interfering task that was inserted in order to simulate the effects of the delay usually involved between seeing a highway sign and acting upon it. The methodology was otherwise the same as for his earlier experiment. In the interference conditions there were more errors for worded signs than for symbol signs, thus suggesting that symbol signs are more resistant to interference than worded signs. In addition, fundamental research on memory would suggest that if a symbol can be remembered as one 'chunk' of information (Miller, 1956) it would be more resistant to forgetting both over time and as a result of direct interference than would the several 'chunks' that may be involved in a worded sign.

In the Ells and Dewar (1979) studies, mentioned above, a later experiment involved presenting the signs in degraded condition by having them viewed through layers of plastic film and by having a glare source in the visual field. Under these conditions the verbal signs suffered more than did the symbol signs.

3.2.5 A symbol used in conjunction with text may be more effective than the text used alone

There is evidence that a symbol used in conjunction with a worded warning can elicit greater compliance than either used alone. Jaynes and Boles (1990) had participants perform a chemistry laboratory task using instructions containing warnings that were verbal only, pictograph only or a combination of both. A control condition involved no warning. The verbal warning was 'WARNING, wear goggles, mask and gloves while performing the task to avoid irritating fumes and possible irritation of skin.' The pictograph conditions involved depictions of all three pieces of equipment. The safety equipment described or depicted was available next to the materials required for the task to be performed. Results showed increased compliance when a pictograph was presented together with the verbal warning (81 per cent compliance), but pictographs alone produced less compliance (34.5 per cent) than the verbal warning alone (63 per cent). The authors point out that the verbal message alone included not only the prescriptive information that the protective clothing should be worn, but also information as to why the protective clothing should be worn. There was no attempt to include in the pictographs information about why the clothing should be worn.

There is a certain logic to the argument that two methods of communication, text and symbol, each catering to partly nonoverlapping sections of the population, should together communicate their message to a larger proportion than either alone. However, the mechanism by which additional conformity was obtained when symbols were used in the Jaynes and Boles (1990) study just described is not clear. They asked participants whether they had noticed and read the verbal warning, and 100 per cent of those in the verbal-only condition claimed to have both noticed and read the warning. Since the warning (with or without pictographs as appropriate) appeared after an introductory paragraph and before the main instructions the 100 per cent figure is not likely to represent hindsight wishful thinking by the participants. Thus, the greater compliance obtained when the symbol was used in conjunction with the verbal warning could not have been due to the symbol's conveying its message to some 'readers' who could not understand the verbal message (participants were all university students), but it must have been a result of the symbol's ability to lend emphasis or salience to the verbal warning.

The study by Otsubo (1988) has been mentioned in Chapter 2. It also used warnings with words only, pictographs only, or both, on a jig-saw (low danger) and a circular saw (high danger) in a behavioural study where participants were asked to use the tool. The warning said, 'CAUTION! SHARP BLADE CAN CUT WEAR GLOVES.' The associated pictograph included two images, one of a circular saw cutting a hand and the second showing a glove being pulled onto a hand. The results relevant to the present discussion, while not significant, were in the same direction as the results of the Jaynes and Boles (1990) study. That is, for the circular saw, when a pictograph only was used, compliance was 22 per cent, when words only were used it was 44 per cent and when both were used it was 50 per cent. This applied only to the more dangerous circular saw. We should note that the pictograph that was used showed what was clearly a circular saw, so participants may have thought the warning did not apply to the jig-saw. In another study, Wogalter et al. (1992) also found a non-significant increase in compliance when pictorials were added to a within-instruction warning.

The difference between behavioural compliance with and without a warning, that is the effectiveness score for the warning, is the ultimate measure we would generally like to obtain. As we have pointed out in Chapter 2, other more easily obtained measures may, as measures of effectiveness, be in some sense second-best proxy measures, but they can give useful insights into the mechanism by which warnings, and symbols, operate. Young (1991) measured the time required to determine if a simulated alcohol label presented on a computer screen contained a warning, when half did and half did not. The presence of a pictorial, warning icon and the use of the colour red for the warning itself all contributed significantly to a decrease in reaction time to determine if a warning was present. Thus the use of a symbol in conjunction with words may serve to emphasise the importance of the words, even though the symbol itself conveys little information.

Young and Wogalter (1990) had participants study one of four versions of an instruction manual for a piece of equipment they were told they would later have to operate. In the manual was a warning that was either in normal print or in more conspicuous print and was either accompanied by a related icon or not. When the warning was in more conspicuous print and also accompanied by the icon the authors found that participants remembered the content of the warning better than in the other three conditions. On a liberal recall criterion 75 per cent were correct in the both-present condition, but only 53 per cent to 56 per cent were correct in the other three conditions. Thus the icon served to draw attention to the warning in a way which enhanced memory for the content of the warning as a whole. In a similar vein, Kline *et al.* (1993) showed that using colour in a warning resulted in the warning label's being perceived as both more readable and more hazardous than when the achromatic version of the label was used.

Another comment on the use of text and symbols together comes from the sphere of human-computer interaction, where pictographs, or icons as they are usually called, are widely used. Gruman (1994) quotes Strijland, who is manager of the human interface design centre at the US AppleSoft division of Apple, as claiming that 'icons coupled with text communicate better than icons or text alone'. An icon on a computer screen may provide cues due to its position, but it is unlikely that the icon does much 'attention getting' of the variety attributed to warning pictographs in the previous paragraph. So how is the claimed increase in communication achieved? It could be that an individual faced with the combination of icon and text may glean some meaning from the icon and some from the (possibly brief) text, thus getting a better overall picture of the meaning than were only text or icon present. It could also mean that some individuals will understand the text (because they do not recognise the icon) and others the icon (perhaps they see it more quickly and simply do not bother with the text). Clearly more research is needed on the precise mechanisms by which symbols, in conjunction with particular contextual cues, communicate their meaning.

Finally, consideration must be given to the purpose the symbol-text combination is being used for. In all the examples given above the effect of the symbol is in some manner to enhance the performance of the whole. The symbol, if considered alone, has always had the same meaning as at least part of the worded message. Research has shown, however, that if the aim of using both together is to educate readers about the meaning of the symbol so that the words can later be removed, then it is unlikely that this aim will be achieved. In the road sign situation Gray (1964) has shown that the symbol component of such hybrid signs is not learned by those who can read. Only 9 per cent of Gray's sample of drivers who drove in an environment using a hybrid 'No entry' road sign were able to identify the symbol alone, but 21 per cent of participants who had seen or studied international signs (where the symbol occurs without words), and 50 per cent of those who had actually driven in a country using the symbol alone, could identify it. Under those international circumstances the users presumably had to ask someone or otherwise actively learn the meaning of the symbol. Thus for those who can read, the symbol may serve a useful purpose in attracting attention to the words in some way, but for those same readers the presence of the symbol together with words that mean the same thing will be unlikely to result in the reader's learning anything about the symbol. Indeed, basic research on what has been called the blocking effect (Kamin, 1969) has shown that when participants (human and animal) learn to respond to stimulus A with response B, and then A is presented together with a new stimulus C (together successfully eliciting response B), stimulus C when presented by itself does not elicit response B, thus indicating that the connection C–B is not learned under these circumstances.

3.3   Problems associated with using symbols

If someone is shown a simple drawing of a railway train and then asked what it is, they will almost certainly reply 'it's a train.' If, however, they are asked what it means the answer will probably depend on the actual or assumed context. If the context is taken to be that of a road sign the symbol might be taken to mean a train crossing ahead or a train station, on a map it could mean a railway line, or in a travel brochure it could symbolise travel by train. Thus the availability of a simple depiction of a relevant object does not necessarily ensure clear communication. This example also serves to emphasise that any testing of the comprehension of symbols should include adequate detail about the context in which the symbol will appear.

   Given that a symbol is well understood in a given context, that advantage will only survive as long as the range of symbols used in that context does not result in the possibility of confusion. If the symbol of a train was used in a railway museum, and there were two types of train, old and new, each symbolised by two appropriately different depictions of a train, there would be some possibility of confusion in the minds of museum visitors. A less hypothetical example is that of a common symbol for lifts or elevators showing two people standing in a box. This symbol, when used on a sign board with an arrow pointing somewhere, is often taken as a symbol for toilets. That may not be an important confusion, but such confusions are possible in safety-related areas unless care is taken to consider the whole range of symbols that might be used in the context under consideration. Lerner and Collins (1983) investigated 18 alternative 'Exit' symbols mixed with 108 building-related foil symbols that included some 'No Exit' signs. The symbols were seen for 4 seconds in conditions that simulated the degradation that would be produced by smoke. Participants had to identify which of the symbols gave exit information. The study found that some of the 'No Exit' foil symbols were interpreted as being exit symbols up to 20 per cent of the time. Thus it would be important to choose clearly-distinct 'Exit' and 'No Exit' symbols for use in a given context.

Another possible problem in using symbols is that their very success in some situations has spawned a tendency to stretch the meaning of well-understood symbols to include situations which they were not designed for. The wheelchair symbol, for example, is intended to indicate access for the disabled, but it is often used to indicate any facility for the disabled. Prohibition is usually indicated by taking a symbol for some activity and putting over it a red annulus and slash. The resulting combination is intended to mean that the activity symbolised is prohibited. Thus if an annulus and slash is placed over a drawing of a piece of beach life-saving equipment (a life-saving reel) the combination should mean something like 'life-saving equipment prohibited here'. Such a symbol was once proposed for use with the intention that it should mean 'No lifeguard on duty.' Appropriate testing would, of course, have revealed the problem.

Sometimes attempts are made to use a single symbol to signify something too complex for any symbol without appropriate learning. For example, the concept, 'In case of fire do not use elevator' is one example which has defied many an attempt to produce a single well-comprehended symbol. Modley (1966) who, together with Dreyfuss (1972), has done much to bring the need for carefully-designed symbols to our attention, has suggested that our present world needs a set of internationally acceptable symbols that can form a system of universally understood visual signs. Bliss (1968) has attempted to produce such a system but it has so many limitations that it has not been used, except in an attempt to assist the learning of the developmentally disabled. An example of a sentence made up of Bliss symbols is given in Figure 3.2. A brief introduction to the Bliss system is given by Dreyfuss (1972).

Although the attempts by Bliss and others to provide for juxtaposing symbols to produce graphic sentences have not been successful, there are

Figure 3.2 An example of a sentence using Blissymbolics. The sentence is intended to mean 'Look up! Low door!'

Figure 3.3   An example of the use of more than one symbol to convey warning information. ('Never leave burning candles unattended', from Zwaga *et al.* 1991).

limited circumstances where, with careful testing, a series of symbols can be used to convey a meaning that is very difficult to convey with a single symbol. Zwaga *et al.* (1991) have produced such compound symbols in the context of instructions for using swimming pool slides (Boersema and Zwaga, 1989) and candles (Zwaga *et al.*, 1991). Figure 3.3 depicts a sequence intending to signify that lighted candles should not be left unattended. In general, where it is difficult to indicate what it is that the observer should do, the meaning may become clearer if what should not be done is also shown. If the drawing of someone sitting beside a lighted candle is seen alone it is difficult to know what aspect of the diagram is important – perhaps the distance between the person and the candle, or the orientation of the person with respect to the candle. The same applies to the second part of the diagram where the empty chair and the cross through the lighted candle make little sense if seen alone. Taken together, the meaning becomes clear. By playing 'spot the difference between the pictures' we have drawn the reader's attention to the salient points.

3.4   The need for testing

There are many examples of research showing that the public often does not understand the meaning that is intended when a commonly seen symbol is used. Some warning symbols that readers may think are well understood, and their test results, are shown in Figure 3.4. The data are from Collins and Pierman (1979).

In the realm of warning symbols, in particular, it is vitally important that the comprehension of any proposed symbol be tested with a sample of prospective users. There is a tendency to assume that if the person who creates or chooses a symbol understands it, and perhaps can convince a colleague that it is a good symbolisation of a proposed referent, then somehow everyone else will either understand it or at least quickly learn what it means. Testing with a

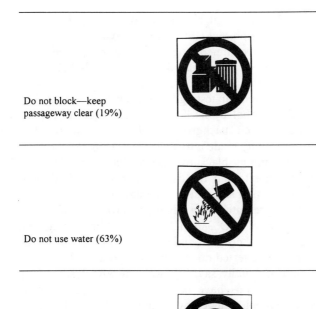

Do not block—keep
passageway clear (19%)

Do not use water (63%)

Do not lock (59%)

Figure 3.4 Examples of poorly understood symbols, from Collins and Pierman (1979). The figures in parentheses are the per cent correct figures the authors obtained from a sample of 143 students, retirees and fire station workers.

naïve population, however, will reveal any problems with the symbol and will also reveal that, even with a good symbol, it may be that only 85 per cent of a sample will be found to understand it. This proportion may be deemed to be sufficiently great, but nevertheless there will, using the criterion suggested, be 15 per cent who will not understand the symbol and the implications of this should be faced.

3.5 Symbol testing

In testing for comprehension of symbols and symbol signs a number of methods have arisen in the course of research over the past half century. These methods have culminated in the development and continuing refinement of international and national standards concerned with the process of standardising symbols for use in a wide variety of circumstances. Methodological details are presented in International Standard ISO 9186 (1989), originally drafted in the late 1970s and at present (1995) in the inevitably slow process of revision. Succinctly presented recommendations relating to sign design, testing and display are also available in Australian Standard AS 2342 (1992). Foster (1994) also provides a summary of testing considerations.

When a symbol is considered for testing, a decision will have to be made about whether to test the symbol by itself, or in the context of a particular sign, with a deliberately-chosen background colour and shape. Warning symbols, for example, are usually made into a sign by setting them on a yellow background within a surrounding black triangle or diamond. The triangle is often used in industrial contexts and the diamond in road-safety or public-information contexts. Prohibition signs have a red annulus and slash superimposed over a black symbol presented on a white background. This colour and shape coding system is part of ISO 3864 (1984). However, testing generally shows that this coding system is not well understood (Cairney and Sless, 1982) in that when participants being tested do not understand a symbol shown in a yellow triangle they do not usually say, 'It's warning about something ....' However, no careful comparisons have been made of either the extent to which placing a warning symbol on a yellow triangle or diamond with a black border actually increases either the proportion of a given population who will correctly identify the symbol as a warning symbol, nor of the conspicuity-enhancing effects that the yellow background and the triangle or diamond shape might be expected to have. The study by Easterby and Hakiel (1977), discussed below, found that for warning signs the precise details of border and background did not generally make any appreciable difference to comprehension rates, but all their tests were with either a thick circular or triangular border. They did not include symbols without any background or border.

When public information symbols are involved, the tendency has been to test a plain black symbol without any surrounding colour or shape, since information-giving signs usually have a black symbol within a white rectangle, or perhaps a white symbol within a plain blue rectangle. Sometimes tourist or national park signs place a white symbol on a plain brown rectangle. Thus the background used for the symbol in the context of an information sign is usually quite plain and would not be likely to affect someone's interpretation of the symbol. Road symbols, on the other hand, are usually prohibitions, regulations or warnings and are usually tested within the conventions of the relevant sign. It would make little sense, for example, to test a 'no right turn' sign without the red annulus and slash. For warnings, the requirement for a particular background and surround is less clear. A 'slippery when wet' symbol, for example, may be tested alone or as a complete warning sign within a yellow triangle or diamond with a black border.

In measuring the extent to which a symbol serves its purpose, a wide variety of methodological approaches has been followed. One of the earliest methods of testing was the brief presentation method of King (1971, 1975), mentioned earlier, which he referred to as testing the 'glance legibility' of the sign. One problem with this method is that it involves choosing a matching stimulus from an array. The results of any such method will always depend on the precise details of the set of alternatives from which the match is chosen. If the matching set includes highly confusable alternatives the success rate will be low. An appropriate use of such a matching method would be where the matching set includes only those signs that might reasonably be confused with the target sign when used in context.

The first set of tests carried out in order to evaluate symbols for the International Organisation for Standardisation (ISO) involved a number of different methods, details of which have been presented by Easterby and Zwaga (1976). The symbols evaluated were for the referents of drinking water, stairs, taxi, waiting room, information and toilet, with from 8 to 35 different symbols initially evaluated for each referent. The large numbers of available symbols required an initial cull done using a ranking method in which participants were first asked to place the symbols into three classes of most appropriate, moderately appropriate and least appropriate, and then asked to rank the symbols within each class. The resulting data were combined using a scaling program to give categorical scale values. Participants from the UK and the Netherlands were used. However, since the data in some cases conflicted, a set of criteria was used to select the three best symbols for each referent for further testing.

The second test asked participants to define the meaning of each symbol they were shown. This comprehension test (originally called a recognition test) involved 50 participants from Argentina, Australia, Canada, India, The Netherlands and Spain. Participants were told, 'We are investigating ways in which we can use signs and symbols instead of words to convey information in tourism, e.g. travel by road, rail or air; using hotels and other forms of accommodation. For each of the symbols we will show you we would like you to name or describe a location or activity you think relates to the symbol on each page.' Participants were shown as an example a sample symbol for a referent not used in the testing. Each symbol was on a separate small (A7) page, with space for a written response. Each subject saw only one version of a symbol for each referent.

The third test was called a 'matching' test and involved 900 participants from each of the above countries, each person providing only a single piece of data for a single symbol. Each target symbol appeared among 23 others, which included one each of the candidate symbols being tested. Thus from an array of symbols on an A4 page each subject was asked to pick the one symbol whose meaning was given in words at the top of that page. There were 18 test sheets, one for each of the three variants for the six referents, with 50 participants in each country responding to each sheet. Thus the test method involved presenting a meaning and asking participants to choose, from an array, a symbol with that meaning. This method perhaps most closely reflects the process applying in a public place such as an airport or bus station where a person knows what they are looking for and sets about searching for the symbol with the required meaning. However, the method is very inefficient in that only one item of data is obtained from each subject. The method also suffers from the recognition-method problem already discussed, namely that the measured success percentages are highly dependent on the distractors used. The method is most applicable to situations where it is important to determine the confusability among a known set of symbols, but it is less applicable to the normal use of warning symbols where one usually sees a warning symbol and then has to work out its meaning. In such a situation the comprehension method is the most appropriate.

The appropriateness ranking and the comprehension methods were used in an extensive study of warning symbols by Easterby and Hakiel (1977). The results will be given in some detail because of the uniqueness of the study and because the original publication is now difficult to obtain. The authors investigated symbols used to indicate a number of hazards, namely fire, poison, caustic, electrical and general hazards. Only two versions of the electrical and fire hazard symbol were available, but there were from 11 to 19 versions of each of the other three referents. All the variants for these three symbols were first subjected to the appropriateness ranking procedure in order to select a restricted range for comprehension testing. The appropriateness ranking results, using 35 observers, were of interest because the symbols for some of the referents included examples of three different types of symbols, namely descriptive ones (those showing the nature of the hazard), prescriptive ones (those telling you what to do) and proscriptive ones (those telling you what not to do). Figure 3.5 shows examples of these. In general it was found that participants preferred symbols describing the hazard rather than symbols indicating what to do or what not to do.

Comprehension testing was carried out with a UK national sample of 4000 participants, each of whom saw one of the four selected versions of each of the five referents. The participants were told, 'I am going to show you some symbols. I would like you to tell me what each one would mean to you if you found it on an ordinary household object such as an article of clothing, a television or a radio set, or any product sold for household use in a bottle, tin or jar.' In addition to the symbols of interest there were distractor symbols commonly seen in the home such as symbols indicating 'Fragile', dry clean only and the woolmark symbol.

The study also had as an aim the determination of the extent to which the background and surround to the symbol would affect its comprehension. Each symbol was presented to one quarter of the participants in one of four versions: a black triangle border on a yellow background, a red triangle border on a white background, a black annulus border on a yellow background or a red annulus border on a white background. Figure 3.5 shows only the triangle version.

Results were coded in terms of the likelihood that appropriate behaviour would follow from the interpretation given, using a six-point scale: certain yes, very likely, likely, marginally likely, unlikely and a final category that was

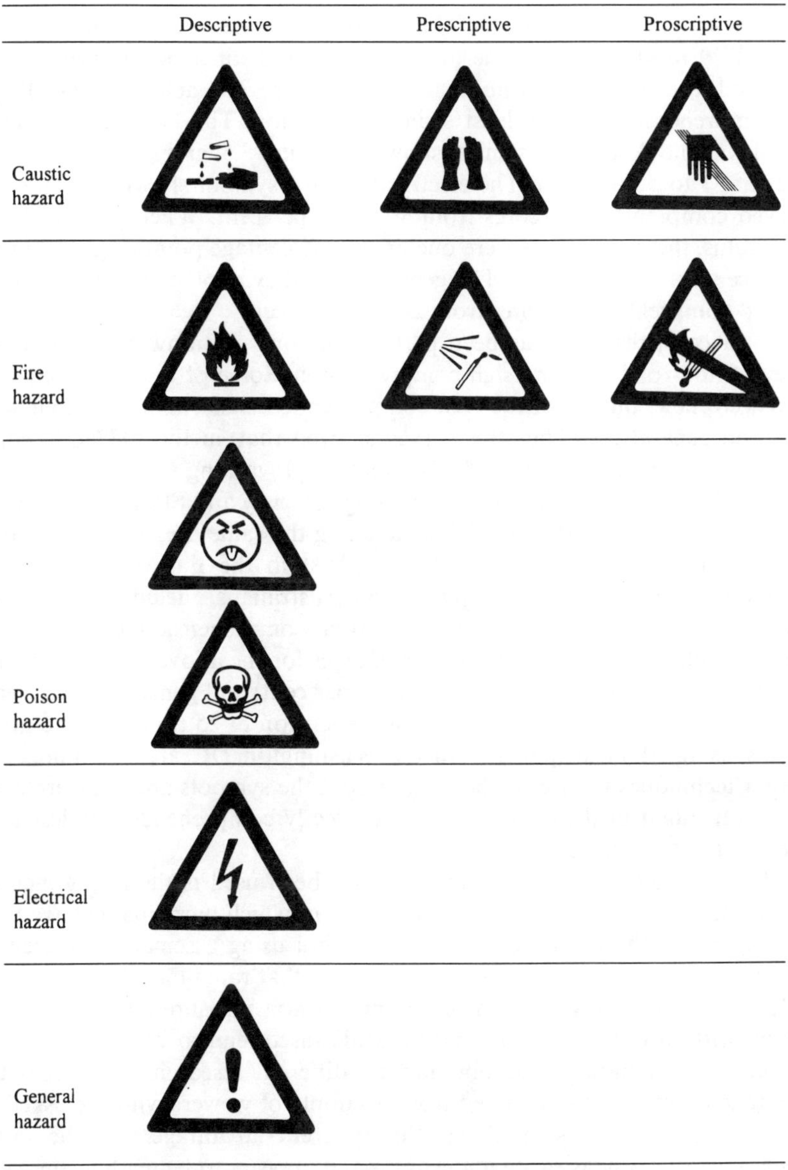

Figure 3.5 Some warning symbols studied by Easterby and Hakiel (1977).

divided into wrong, opposite or don't know responses. Overall, the borders and backgrounds made no clear differences to comprehension rates. The best fire symbol, the descriptive symbol shown in Figure 3.5, gave comprehension scores of from 53 per cent (certain yes + very likely) to 60 per cent (all categories from certain yes to marginally likely). Using these same categories the best poison hazard symbol, the skull and crossbones, gave comprehension scores of 29 to 48 per cent. The only graphically distinct poison symbol that was tested, the 'Mr. Yuk' symbol (shown in Figure 3.5), achieved only 19 per cent comprehension on the least stringent criterion. The best caustic hazard symbol was the descriptive symbol shown in Figure 3.5, giving comprehension scores of 51 to 55 per cent. The electrical hazard symbol shown in Figure 3.5 achieved comprehension scores from 40 to 44 per cent. When presented in a red annulus, the percentages were one or two percentage points higher, but the difference was not significant. The general hazard symbol shown in Figure 3.5 achieved comprehension scores from 24 to 39 per cent.

This piece of research has been perhaps the only nationwide large-sample survey of non-road warning sign comprehension. Some of the signs that were tested were new, but all of the best-comprehended signs were ones that were already in general use. Thus this survey showed that, in the 1977 UK population at large, comprehension of commonly-used warning signs was not high. Other studies have corroborated this finding. Cairney and Sless (1982) tested a series of occupational safety symbols, including the same fire, poison, electrical and general hazard symbols found to be best in the Easterby and Hakiel (1977) study. They tested a sample of 154 Australians, including many new immigrants studying English. Using a strict comprehension criterion they found the following comprehension percentages for the above symbols: fire (26 per cent), poison (25 per cent), electrical (29 per cent) and general (21 per cent). Collins and Lerner (1982) tested the comprehension of 25 proposed fire-safety symbols using 91 participants from the Washington DC area. A number of different techniques were used, both to present the symbols and measure comprehension. Eight of their symbols were correctly comprehended by less than 25 per cent of their sample.

A distinctly different test method that has been used is the rating method (Dewar and Ells, 1977; Green and Pew, 1978), in which participants have been required to rate the meaningfulness of a symbol using a semantic differential measure. However, the studies referred to found that rated meaningfulness was highly correlated with definition or comprehension measures. There is also a problem with rating methods in that the scales used tend to vary, thus making it difficult to compare results obtained by different researchers, whereas the percentage correct figure obtained from a sample of viewers who are asked to give the meaning of a symbol has the apparent advantage that the single 'percent correct' value is easily understood and appears to be readily comparable between different studies. The fact that differing scoring criteria for correctness may be used is often not given the consideration it deserves, but nevertheless the easily comprehended percent-correct measure is now almost the only index of comprehension that is used. Collins and Lerner (1982) highlight problems that can arise due to differing definitions of 'correct' in a comprehension measure.

The final method to be discussed is the recall method. There are some situations where a pictorial or concept-related symbol is difficult if not impossible to produce. In such situations, where an abstract or arbitrary symbol may be the only possibility, it would be unlikely that a comprehension test would produce even a small percentage of correct responses. In such circumstances a recall test is an appropriate measure of the symbol's performance. Walker *et al.* (1965) used a memory test for assessing roadway signs and Cairney and Sless (1982) go so far as to suggest that, 'Since most public information symbols will be used in situations where people encounter them repeatedly, ease of learning would seem to be an important criterion to consider in selecting a symbol ....' We discuss their method more fully below.

These early testing studies served two purposes. They served to refine the details of the testing methods now incorporated into the international standard ISO 9186, and summarised below. They also make it clear to all who have been involved in the development of testing methods that it is difficult to persuade purse-string holders of the need for testing symbols, and that funds for testing are therefore hard to obtain. The grand-scale study by Easterby and Hakiel (1977), for example, has been the only well-funded study outside the area of road signs. Testing must therefore be done in the most efficient way possible. What follows is an outline of the present and proposed methods used in the ISO standard on testing (ISO 9186), together with recommendations for testing when resources are limited. A series of steps is involved, only some of which may be appropriate in any given situation.

3.5.1 Is the symbol needed?

Before we embark on testing something like a symbol we should first ask whether it is needed. What behaviour is happening without the symbol, for example, that we would reasonably expect to change if one were used? The reason for such a consideration is that symbols are often used without any serious thought as to their likely effectiveness. With public information signs, local government bodies often feel compelled to put up warning signs and the like simply to cover themselves against prosecution. They do this even when it is highly unlikely that the sign will change behaviour. For example, one council in the Sydney area felt they should warn swimmers: DANGER − RIVER SUBJECT TO SUDDEN RISE − WATER CAN BECOME VERY COLD. What does 'sudden' mean and how quickly could the water become cold? Is it likely that such a sign will change behaviour in a popular swimming place that has been used for a long time without any obvious problems? Our earlier discussions have stressed the distinction between compliance and effectiveness scores. In considering the potential effect of the use of a symbol it is the potential for increased compliance, that is, the potential effectiveness or impact that should be stressed. As we have discussed in Chapter 2, this requires that data be obtained both with and without the item of concern, in this case the symbol.

3.5.2 Appropriateness rating

Once it is decided that a symbol is needed there may be a number of alternative symbols available. The simplest method of choosing the best of a series of alternative symbols is to have a number of people who are unfamiliar with them independently rate them in order of suitability. The suggested method (Foster, 1991) is that three categories are used for rating each symbol: highly appropriate – most people will understand it; slightly appropriate – only about half the people will understand it; and not appropriate – almost no one will understand it. The idea that independent raters can fairly accurately judge the proportion of those seeing the symbol who will understand it comes from the work of Zwaga (1989) who found that such ratings are highly correlated with subsequent comprehension measures. In particular, Zwaga found that if raters judge that a symbol will be comprehended by a very high proportion or a very low proportion of the population they are likely to be right. Judged proportions in the central ranges will require confirmation with comprehension tests. The precise criteria that are used will depend on the balance between the costs of symbol misinterpretation in relation to the benefits of correct symbol comprehension – all considered in the light of the cost of further symbol development and testing. Where a symbol is being used to provide public information of a non-emergency nature, Foster (1991) has suggested that if 65 per cent of a group of 50 or more respondents judge that a symbol will be 'highly appropriate – most people will understand it' then testing need proceed no further and the symbol can be accepted, subject to a consideration of design details from the point of view of legibility and confusion with other symbols. Otherwise, further testing should be carried out on the three graphically distinct symbols that the appropriateness-rating process has suggested will be the most suitable. If fewer than, say, 20 raters have been used then it would be advisable to proceed to a comprehension test in any case. In the warnings domain these criteria would be appropriate for a warning-related symbol which is designed primarily for its attention-getting or alerting properties and for which supporting textual information is also provided.

3.5.3 Comprehension

The symbol testing method that is most easily understood and accepted is the comprehension method, that, in brief, involves showing suitably chosen participants the symbol and asking what it means. Of the testing methods available it is the one most clearly applicable to situations involving warning signs, where in normal use one sees a warning sign and has to determine its meaning. However, the process of carrying out a comprehension testing session requires some thought. Participants, for example, should be representative of the ultimate users of the symbols and they should be naïve to the extent that they should not have been involved in any way in the symbol development. On the other hand, they should certainly be familiar with the concepts being symbolised. If a number of symbols are to be tested together, each person should be asked about only one candidate symbol for each referent being tested. The simplest testing procedure involves making up booklets of symbols, in several different random orders, each booklet involving only one of the candidate symbols for each referent being tested. Participants should be told the context of the symbols with a statement such as, 'I am going to show you some symbols that might be used on the packaging of products you might buy for the home. Please write what you think each symbol means on the line below the symbol.'

One problem with comprehension testing is the scoring of results. A number of judges should be involved – the ISO standard requires three. The responses should be scored as follows (Foster, 1991):

1 The response given is 'don't know'
2 The meaning which is stated is the opposite to that intended
3 Correct understanding of the symbol is certain or likely
4 Any other response
5 No response is given.

The next difficulty is to decide on the criterion for acceptable performance. Should 100 per cent of those shown the symbols get it correct (Category 3), or would 66 per cent be sufficient? Various criteria are in use, ranging from 66 per cent in the International Standard (ISO 9186–1989), where it is assumed that respondents will be from a variety of countries, to 85 per cent in the relevant US (ANSI Z535–1987) and Australian (AS 2342–1992) standards, where it is assumed that the respondents are from a single country. Another proportion of interest is the proportion of people who give an opposite response. In safety related situations even a small proportion of such responses should be cause for concern.

3.5.4 Learning

Finally, if a symbol is to be developed for a totally abstract concept, or if reasonable attempts at choosing a suitable symbol that may be recognised have failed, then a symbol should be chosen on the basis of its being easily remembered. If you tell people what it means one week, can they remember the next? A learning test is under consideration for inclusion in a revised ISO 9186 (Foster, 1991) and one is described in detail in Australian Standard 2342. A problem with using such a test is that if only one symbol is shown then the learning test will almost certainly show a high rate of recall. To make such a test representative of real life any symbol under test should be part of a series of at least eight symbols. If fewer than eight symbols are of interest, then additional symbols from the same general contextual area should be used to increase the number to at least eight. Participants are first asked to guess the meaning of each symbol. They are then told the meaning and, without warning, are asked a week later to give the meaning again. The final measure is then a per cent correct, but as it is taken a week after participants are told the meaning of the symbol it is held to be a measure of ease of learning.

3.6 Designing and using symbols

Up to this point nothing has been said about the design details of symbols, nor of practical matters such as the consideration that should be given to the possibility of confusions arising among a series of somewhat similar symbols, nor about the process of urgency mapping with symbols so as to relate the symbol in some way to the severity of the situation to which it related. Often symbols will not be designed from scratch but taken from already published sources that include only symbols of relatively good graphic design, such as AIGA (1981), Dreyfuss (1972), FMC (1985), ISO 7001 (1990), Modley (1976) and Westinghouse Electric Corporation (1981). But few of the symbols in these sources have been subjected to any pre-publication tests for comprehension. For example, Mayer and Laux (1989) found when they tested 16 pictograms from the Westinghouse Product Safety Label Handbook (Westinghouse Electric Corporation, 1981) that only three achieved an 85 per cent comprehension criterion. When a symbol is taken from sources such as those quoted above it is probably of a relatively high graphic design standard, but the extent to which it will be comprehended in any given context is by no means clear. When the symbol is designed from scratch, there are even more matters to be considered.

3.6.1 Legibility

When a symbol is designed from scratch, careful consideration will have to be given to legibility. Modley (1966) and Dreyfuss (1972) give much useful guidance on theoretical approaches that will assist the designer in the creative process of developing a design concept and then refining the design concept into a useful image. The next step is to ensure that the final design does not include such fine detail that it will be difficult to see if it is small, far away or if the light is poor. One way to assist legibility is to design the symbol on a 20×20 grid and not to make any significant detail smaller in size than one square unit. This and other design recommendations are given in ISO 7001–1990 and in AS 2342–1992.

Some recent work by Kline and Fuchs (1993) suggests that it may be possible to be more scientific about the process of ensuring the greatest possible legibility distance. They have argued that symbols can be made more visible by reducing the spatial frequencies involved and they give some road sign examples. They were able to increase legibility simply by removing high-frequency spatial changes in detail. To oversimplify the process somewhat involves equalising as much as possible the sizes of figure and background details within the symbol. This is a fascinating development that we hope will be taken further.

When text is included, either within the symbol sign or even in the surrounding text, matters of text size, text and background colours, typefaces, leading, white space and the whole world of typographical detail will affect the extent to which the symbol or symbol sign will be legible. To the extent that those matters are understood they will be left to the existing literature (e.g. Easterby and Zwaga, 1984), but many of them could well be subjected to more rigorous testing, as discussed below, particularly when novel combinations or layouts are used to convey specific information.

3.6.2 Conspicuity

Often one of the reasons for using a symbol on a road sign or a product warning is the desire to have something which will stand out from the surrounding background. Indeed, research on the use of symbols in conjunction with associated wording, discussed earlier, indicates that one way in which the symbol-plus-words combination produces its enhanced effect is through the attention-getting effect of the symbol. Hence an important property of a symbol is its conspicuity. However, Cole and Hughes (1990) have pointed out that conspicuity can apply in two situations: when the observer is not expecting anything in particular and when the observer is actively searching. They refer to the former as attention conspicuity and the latter as search conspicuity. In relation to a warning sign on a product, both of these aspects of conspicuity are important. A warning symbol sign should be able to command attention when we are not looking for it. On the other hand we may be looking at a product label for information about safety-related matters, perhaps with an image in our mind of a particular warning sign we are looking for. That symbol, if present, should therefore have high search conspicuity.

A reason for being clear about these two aspects of conspicuity is that each implies a different measure. If attention conspicuity is to be measured it is important that the subject not know what it is that they are being asked to report. For example, a measure of attention conspicuity described by Cole and Jenkins (1980) is to flash the relevant symbols briefly on a screen, say for 200 m s, and ask observers to report what they see. More conspicuous items should be correctly reported more often.

Search conspicuity, on the other hand, is measured under conditions where the subject knows what he or she is looking for and has to report if it is present. A measure of search conspicuity would be the time it takes participants to report the presence of a known symbol, perhaps with backgrounds of varying degrees of distraction. Such a measure was used in a study by Boersema *et al.* (1989) where eye movements were used. The study showed that nearby advertisements can distract attention from a sign. Using slides of railway station scenes that included a routing sign, the number of advertisements that were visible near the routing sign was artificially manipulated. An eye movement camera enabled a continuous record of when and where participants directed their gaze. Participants were told before each slide what routing sign to look for. In general it took longer to find the routing sign when there were more distracting advertisements.

One might argue that the absolute times recorded in laboratory studies are unrealistic, but the point of such laboratory studies is to compare conditions rather than to estimate real-world response times. In the Boersema *et al.* (1989) study, what mattered was that the search time was longer when distracting advertisements were near the informational sign. On the other hand, there have also been a number of studies showing that laboratory measures do relate quite well to real-world measures. With respect to conspicuity, see for example Hughes and Cole (1986).

A measure of conspicuity that requires only a few participants was used by Adams and Montague (1994) in a study of warning signs to be used at the side of railway tracks to warn train drivers of track works ahead. A number of alternatives were compared by mounting each one to scale on a separate A3-sized photograph of a realistic railway track scene. An identical photograph was prepared without any sign on it. The required photograph was set up, at eye height, at the end of a 25 m walkway. There was a series of trials and on any one trial there was a 50 per cent chance that the displayed scene would include a sign. Observers were given a card with the sign drawn on it and asked to walk towards the display until they could say if the known sign was present. The experimenters argued that a more conspicuous sign is one for which the train driver would be prepared to say, 'I see something there' from a greater distance.

3.6.3 Discriminability

If a symbol sign is to stand out at a distance, both the legibility and the conspicuity considerations already discussed are important. But there is a further consideration. A symbol might stand out, yet because the observer may have in mind two or more possible symbols that look the same and appear in that same context, he or she will take longer, or approach closer, to be sure that the correct symbol has been identified. Thus a warning symbol sign should be highly identifiable – should stand out clearly – from all possible alternatives. Foster (1994) discusses a wide range of measures of discriminability, but perhaps the simplest is first to show naïve participants a display of all the symbols that may be seen in the context under consideration. Participants are then given a sheet with a list of all the possible meanings that could apply in that context and shown the symbols one at a time with the request that, as each symbol is shown, they should mark on the sheet the meaning applying to each symbol. They are not told that there is only one slide for each meaning. Any possible discriminability problems will show up as confusions among the symbols.

In the study by Adams and Montague (1994), discussed above, discriminability was also measured. For conspicuity the question was 'Is something there?' For discriminability, however, the question was, 'What is there?' The discriminability question of concern in the study was which of two pairs of signs was the more discriminable. Subjects were given a card with both of the signs from one of the sets on it, each marked with a letter. On the photograph at the end of the walkway was one of the signs and participants were asked to walk towards the photograph until they could say which sign was displayed. This procedure occurred separately for each of the two sets of signs being evaluated. The results showed that for one set the participants had to approach to a much closer average distance to be able to say correctly which sign was there, thus indicating that the two signs of this set were less discriminable from each other than the two signs of the other set.

3.6.4 Urgency mapping

In the literature on the design and testing of warning symbols nothing, to our knowledge, is mentioned about the necessity of relating the urgency of the symbol sign to the severity of the situation signified. Sometimes with road warning signs a particularly hazardous referent will be signified with a particularly large or conspicuous sign, but in general the unexamined assumption is that the sign is there to ensure 100 per cent compliance. With worded signs, as already mentioned, there has been research on the relative urgency of signal words and colours, but with symbols there is so much to deal with in simply ensuring that the symbol is understood to mean the same as its worded counterpart that other aspects have been ignored. The symbol itself (as opposed to the background colour and the surround shape) is seen as a replacement for the informational part of a worded warning. Symbols are used to mean things like 'Electrical hazard', 'Poison', 'Keep hands away', 'Corrosive'. In all these situations the intention is that the symbol should be a direct replacement for the relevant words and hence should provide the same informational content.

A warning symbol sign, however, usually involves more than just the symbol. As already discussed, the warning symbol almost always occurs in combination with other graphic elements. For example, the symbol usually occurs on a background of a particular colour and within a surround of a particular shape. To some extent these other symbolic elements replace the signal word that might be used in the words-only version. Thus, 'Warning, electrical hazard' will be replaced by a black symbol that will be presented on a yellow background with a heavy black triangular or diamond border. The implication is that the border and background represent the 'Warning' part of the complete symbol sign and that the icon represents the 'Electrical hazard' part.

If this distinction is appropriate we should be able to develop a scale of urgency appropriate for warning signs by scaling the relative urgency of a number of alternative background colours and shapes. There has, as discussed in Chapter 2, been work on the scaling of colours, although there has been no systematic attempt to put that scaling information to work in graded signs. In at least one jurisdiction red has been associated with the word 'Warning' in worded signs and yellow with the word 'Caution', but concern over the confusion between the relative hazardousness connoted by the words 'Warning' and 'Caution' has overshadowed any consideration of possible apparent hazardousness differences that may exist between the associated colours.

There has been at least one attempt to scale shapes. Riley et al. (1982) chose a series of shapes that might be used for the surround of warnings. These included the square, circle, triangles in various orientations, various rectangular shapes and a number of polygons. The aim of the research was to choose the best shape for warnings, without any consideration of the possibility that the various shapes studied might be used for warnings of varying degrees of urgency. Participants were asked to judge all possible pairs of the 19 chosen shapes. In each case they were asked to indicate, '. . . which shape of each pair represented the preferred warning indicator (i.e. the one that would be more likely to attract their attention)'. The results were scaled and the best shape was found to be the point-down triangle, followed by the diamond and then the octagon. A finding of interest was that a clear distinction was made between the various orientations of the same shape. Thus the point-up triangle orientation was rated considerably lower than the highest-ranking point-down triangle. The authors consider that their results reflect our learned associations with these shapes, mainly through the road-sign conventions.

Thus there has been some work that could form the basis for systematically relating the urgency of the various parts of a symbol sign to the degree of risk or hazard associated with the referent. That little relevant systematic work has been done is perhaps because, unlike the auditory situation, there is no obvious set of dimensions such as loudness that are naturally related to urgency. Studies such as Adams and Edworthy (1995), Braun and Silver (1995), Chapanis, (1994) and Kline et al. (1993) have shown that the urgency-imparting effects of aspects of visual signs such as colours and line weights can indeed be quantified.

There is a further issue that needs to be considered in the urgency mapping of symbol signs and that is whether there is any interaction between background colour and surround shape in determining the the perceived urgency of the whole. On top of this, there is the possibility that some symbols have an inherent alerting effect as well as an informational content. For example, the electrical hazard and the poison symbols shown in Figure 3.5 almost certainly have some alerting properties. Thus it is possible that colours, shapes and images cannot be considered in isolation and that any scaling that occurs will have to be of the final symbol sign as a whole.

That there is a need for appropriate matching of the urgency of symbolic warning material to the requirements of various referents is evidenced by a survey carried out by the Australian Department of Health (1991) where householders were asked to assess the likely outcomes of a series of accidental poisoning scenarios involving some 20 household products such as superglue, weedkiller, and Valium tablets. They made these ratings both before and after seeing a package of the relevant product. The results were complex, but there were many misconceptions about the seriousness of the scenarios presented, including many overestimations of the seriousness of the likely consequences. However, exposure to the packages of the relevant products at the time of the interview did not change the accuracy of the ratings, nor was it true that those who usually kept the product in their home were better at judging the seriousness of the scenarios' consequences. From this research, at least, it appears that existing warnings are not particularly effective in conveying information about the relative danger of various household products. In one case, a package had the word 'warning' writ very large and a householder commented, '. . . they've put it in such big letters on that pack that they must mean it'. One can imagine the same comment being given to a very large symbol on a package. Thus there is at least some implicit understanding in the minds of the public that it is possible to relate the urgency of warning material to the severity of the situation to which it applies. All the more reason, therefore, to test not just the comprehension, but also the overall urgency, of warning symbols, and indeed warning material of any sort, in the context in which it will be used – that is, on the package.

3.7   Testing warning symbols in context

As indicated, it is most important in evaluating warning symbols that consideration is given to testing the final complete package or presentation that incorporates the symbol of concern. It is one thing to test the comprehension of a symbol when someone is told 'here is a symbol you might see on a package ....' But the final product may be so designed that the symbol is only seen in conjunction with particular instructions, or only after turning to a particular page, or perhaps after opening a particular cover. Worse still, the symbol may be on outer packaging that is discarded and so never seen close to the time at which the product is used, or the symbol's colours may be such that it is difficult to see in normal use conditions. These details of the circumstances of use may well affect the extent to which the symbol assists in achieving compliance with the warning of which it may be a part.

As we have discussed earlier, a warning provides only one source of information, one input, into the complex process of deciding whether to perform the behaviour that we would regard as demonstrating compliance. In this regard, a warning symbol is no different from the written warnings discussed in Chapter 2. Consider, for example, a 'No diving' warning symbol sign placed at the shallow end of a swimming pool. Such a symbol sign is very readily comprehended (AS 2342–1992), yet there has been research showing that when this sign was added at a high school pool the incidence of diving at the shallow end increased for some user groups (Goldhaber and deTurck, 1989). It appeared that many users had dived safely at that pool prior to the placement of the sign and hence the sign presented, to school children, something of a challenge. Thus the effectiveness of the sign, in context, was negative. The context in which any warning is given will always affect the extent to which the warning is complied with, but with symbols there is a tendency to assume that somehow the use of a symbol has added a new dimension that will magically improve compliance. Such is not the case. The overall package, or context of use, needs to be considered in determining the effectiveness of the symbol-in-context.

CHAPTER FOUR

Cautionary tales: problems and solutions for auditory warnings

4.1 Introduction

Arenas where warnings are typically used in great numbers are high workload, relatively high-stress working environments. The types of warnings normally used in such environments are those which require an active response within a relatively short period of time. Such situations lend themselves to auditory alarms where the active cancellation of the alarm by the operator is an integral part of the normal alarm handling process. Auditory warnings are also used in such situations for a number of other reasons: they are useful if the operator's visual field is already full, if the operator is some distance from the source of the problem, and if the environment is noisy in such a way as to make verbal warnings hard to detect. Such environments include the flight deck of an aircraft, the factory floor and the hospital intensive care unit.

With verbal warnings there is some dissociation between the warning itself and the associated instructions, as we have seen in earlier chapters. With auditory warnings that dissociation is particularly clear. Auditory warnings are the artefacts which draw attention to a problem, whereupon the operator will then look for further information. If warnings are well designed and a training programme is entered into, the meanings of the warnings can be learnt and auditory warnings can impart information as well as simply serving as alerts. The issue of urgency mapping is therefore more obvious, but also more directly addressable, in the case of auditory warnings. Much of this chapter is aimed at showing how the issue of urgency mapping has been approached in practice, although many other important problems associated with auditory warnings are also considered.

Just as for visual warnings, the temptation with auditory warnings is to overuse them. Designers tend to operate with a 'better-safe-than-sorry' philosophy which dictates that if there is any doubt, a warning should be implemented. This may be technically admirable, but it may not be psychologically desirable, as the operator may become overloaded and thus made relatively ineffective. The auditory channel is more prone to overload than the visual channel because it does not have the processing capacity, in bits per second, of the visual channel, so the issue of overload is more pertinent. Equally, we find it hard to deal with many sounds at the same time, whereas with visual warnings we can control the temporal patterning of our observations (which may, of course, be our downfall). Therefore in some ways it is even more important to have some sort of rationale for the implementation of auditory warnings than for verbal warnings. The issue of compliance and effectiveness scores is equally as relevant for auditory warnings as it is for other types of warning (and in some ways it is easier to measure, because of the way auditory warnings are typically used). However, we believe that many design issues are still to be explored so we will focus on these in this chapter. The problem of design will be approached by highlighting two areas for which design solutions have been proposed and to some extent implemented. The first area encompasses intensive care units and operating rooms of hospitals, and the second is that of the helicopter cockpit.

4.2 Hospital alarms

4.2.1 Alarm use in Intensive Care Units

There is very little objective evidence concerning the way alarms are actually used in the intensive care unit, but there is some data from questionnaire studies. A study by McIntyre (1985) showed that a large sample of anaesthetists generally favoured the use of auditory warnings in the operating theatre. The survey, which was carried out by a questionnaire sent to a large number of anaesthetists, confirmed that most of the equipment used by anaesthetists is equipped with alarms. The data also indicated that auditory warnings often do warn anaesthetists of potential problems.

Table 4.1 shows the numbers of responses obtained in relation to questions about alarm activation and deactivation. These responses demonstrate generally good practice with reference to alarm activation. For example, the majority of responses to the question 'When you use an instrument that incorporates an audible alarm device do you routinely activate the alarm system when you begin to use the instrument?' was 'Yes'. Such practice not only checks that the alarm is working properly, it also reminds the anaesthetist of the warning sound itself, which will improve the likelihood that it is recognised during anaesthesia if it sounds. The survey also shows that the majority of the respondents do sometimes deliberately deactivate the alarms. There are many reasons for this, as Table 4.1 shows. The occurrence of many false alarms is the most popular excuse, followed by the need for peace and quiet, followed by dislike of the actual alarm sound itself. Thus, if the effectiveness of hospital alarms is to improve, perhaps the most basic design criterion should be to design alarms in such a way to ensure that they are

Table 4.1 Answers given by anaesthetists to questions about alarm activation and deactivation (from McIntyre, 1985)

'When you use an instrument that incorporates an audible alarm device do you routinely activate the alarm system when you begin to use the instrument?'	'Yes' 565	'No' 224
'Have you ever deliberately deactivated an audible alarm device at the start of a case?'	'Yes' 460	'No' 339
'Why do you sometimes/always deactivate an audible alarm in this fashion?'	Number of positive responses	
Need for peace and quiet	119	
Desire to keep information to yourself	45	
Consider the existence of the alarm hinders you from looking after the patient properly	45	
Too many false alarms during a case	314	
Confusion between different audible signals in the operating room makes the particular alarm useless	53	
Dislike the character of the particular sound	79	

switched off less often. The problem of false alarms is perhaps an engineering problem that will need solving through a somewhat different route, but if alarms can be designed to be more tolerable, the anaesthetists' apparent aversion to false alarms may be reduced, with the result that because the alarms are less intrusive they will be more likely to be left turned on.

The anaesthetists were also asked which of the characteristics of the alarm enabled them to identify correctly the source of the warning signal. The responses to this question provide some insight into the way in which alarms are typically used. The results are shown in Table 4.2, where three clear functions of the alarms are shown. The first is in the identification of the problem through the distinctive sound of the alarm itself; the second is the use of an auditory alarm to direct attention to a visual display; and the third is to indicate the direction from which the alarm is coming. These results certainly make sense in terms of what we would generally expect auditory warnings to do, and thus how they should be designed. In particular, knowing the meaning of the warning is important, because the hearer will then be able to identify the unique problem which the alarm is signalling.

One aspect of the results which is slightly surprising is that the location from which the alarm is coming was not used to a greater extent. This, we believe, is largely because the acoustic structure of the alarms does not allow such localisation – more sophisticated warnings could give very useful information concerning location and the ears of the hearer would automatically use

Table 4.2   Reasons given for recognition of auditory signals (from McIntyre, 1985)

When an audible alarm activates, which one of the following
most enables you to identify the source of the signal

	Number of positive respondents
Recognise the distinctive sound of that alarm	422
Recognise the direction the sound is coming from	43
Look for an associated visual indicator	172
Other	0

this information in locating the source of the sound. However, we will come to this later. We should also remind ourselves that McIntyre's (1985) useful study is based on retrospection and is subjective in nature. A study by Momtahan *et al.* (1993) shows us that what people believe about their ability to recognise alarms, and their ability actually to do so, are not necessarily the same thing. A further and very revealing aspect of McIntyre's study is the response to a question about the number of alarms with which anaesthetists thought they could cope. The results are shown in Figure 4.1. Here the respondents are showing good insight into memory capacity – the most popular number, by a long way, was five, and the vast majority of responses was lower than this.

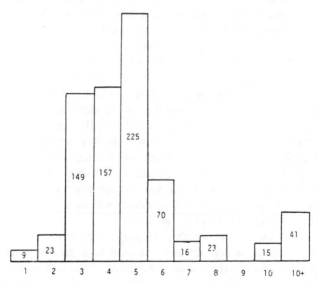

Figure 4.1   Number of individually recognisable alarms anaesthetists say they can cope with (from McIntyre, 1985).

Other, objective studies (e.g. Patterson and Milroy, 1980) have also suggested that alarm numbers should be kept to a minimum.

Thus McIntyre's study shows that auditory alarms are useful to anaesthetists but that anaesthetists have some misgivings about them. We should add that other information from McIntyre's study makes it clear that the possibly inappropriate behaviour engaged in by anaesthetists with respect to alarms is indeed a function of the alarms and their operation and is not due to perverseness on the part of anaesthetists.

A related issue which will have an impact on the effectiveness and use of auditory warnings in the hospital environment is the frequency with which alarms sound, and their perceived importance. A study by O'Carroll (1986) required nurses to record on paper the origin and frequency of all alarms sounding in the intensive care unit over a three-week period. The results of this study indicate that, of 1455 alarm soundings, only eight represented life-threatening situations. There were clear differences between the severity of individual incidents, as one would expect, but it seems that very few were life threatening. O'Carroll's main suggestion is that there is a need for a graded system of alarms in order to cope with the graded severity, or urgency, of the situations which the alarms signal. O'Carroll (1986) is thus advocating some system of urgency mapping. What is certainly not needed is the present series of unrelated, shrill, often unrecognisable alarms all demanding attention and none giving precise cues, at least to the first-time listener, as to the nature of the problem or as to how quickly the listener should be reacting.

4.2.2 Specific alarm problems

In this section the main problems associated with hospital alarms will be described in detail. These problems are that warnings are often too loud, there are too many of them, that they are confusing both because of their number and their acoustic qualities, and that there has been, until recently, no real attempt to integrate warnings into the work environment and relate warning sounds to individual medical problems. There is also a big problem with false alarms, which is a largely an engineering problem and will not be discussed at length.

4.2.2.1 Number

The problems associated with the sheer number of warnings in certain hospital environments are at once the most obvious and the most intractable of the issues which need to be addressed. Studies have found more than 20 alarms associated with a single patient (Meredith and Edworthy, 1994), and Momtahan et al. (1993) found a total of 49 alarms in one operating theatre and recovery room. It is relatively easy for us to identify hundreds of sounds in our normal acoustic environment, but this ability does not, it seems, transfer to the identification of a multitude of abstract sounds which may hold little or no meaning for us.

Perhaps the biggest problem regarding the number of warnings in the environments we have been discussing is the increasing tendency of equipment manufacturers to have their equipment generate an auditory warning whenever possible, simply because the technology allows and even seems to invite it. Perhaps the threat of litigation if an accident were to result from the lack of a warning where one was possible is at the back of every manufacturer's mind. Whatever the reason, a result of the proliferation of warnings is that it is not unusual to find a single patient in an intensive care unit linked to several pieces of monitoring equipment which, between them, can produce 20 or 30 different warning sounds. When so many alarms are present, medical staff are clearly not going to be able to identify individual warnings. This inevitably leads to alarms sounding and not being recognisable. Fatalities in anaesthesia, for example, have been attributed to confusions between leads which produce warning sounds when disconnected (Cooper and Couvillon, 1983). Although this may not be one of the most important causes of fatalities, compared with other human and technical factors, the sheer number of auditory warnings does cause many problems for medical staff.

A study by Momtahan and Tansley (1989) demonstrates a further problem of overconfidence in ability to identify warnings. In this study, the alarms in the recovery and operating rooms of a large Canadian teaching hospital were tested for recognisability. Momtahan and Tansley found three important general problems. There were too many alarms to be remembered successfully; some alarms were similar enough to be confused with one another; and some had the potential to mask one another. Medical staff – both nurses and anaesthetists – were assessed on their ability to identify correctly the 23 alarms in use in the recovery room at the time of the study (another 26 functioned in the operating theatre). Nurses correctly identified an average of eight recovery room alarms and the anaesthetists correctly identified an average of five. Similar proportions were found for the operating theatre alarms. Results showed that the anaesthetists significantly overestimated the number of alarms that they believed they could identify correctly, showing that people's self-reported facility with alarms may well be overestimated.

Thus, field studies have shown enormous real-life difficulties with situations involving a large number of auditory alarms – and we would add that the problem of increasing alarm numbers applies in the aviation environment (Doll and Folds, 1986) as well as in the hospital environment. A laboratory study in this aviation context lends further support to the problem we have been discussing. Patterson and Milroy (1980) asked participants to learn a set of civil aircraft warnings on a cumulative basis. They found that as the number of warnings increased up to about five or six they were easily learned, but beyond this number additional warnings appeared to be much harder to commit to memory. It is a well-established fact that our memory for unrelated, arbitrary items of information ranges from five to nine items (Miller, 1956). As it is unlikely that cognitive processes will change much in the near future we should attempt to rectify the problem by having technology acknowledge our cognitive limitations rather than by expecting our mental capacity somehow to catch up with technology.

To the present unfortunate situation there are two possible solutions. One is to use our knowledge of human information processing to devise a radically different set of alarms which are easier to learn. This alternative, which is considered further in Chapter 5, might increase our capacity slightly, but nothing like the amount necessary to cope with the 30 or so warnings some environments at present contain. The other alternative is to reduce the number of alarms. Reducing the number of alarms, however, is something which is not simply done. If many of the present warnings were removed simply on the basis that they were not essential, leaving only the most critical ones, it would inevitably be argued that potentially critical events might be missed. If the number of warnings is to be reduced, then reduction must be achieved in some systematic, rational way. Fortunately there have been some attempts to reduce numbers systematically, through research and the development of new standards. We will discuss them later.

4.2.2.2   Deactivation of alarms

There is no doubt that users of medical equipment sometimes deactivate alarms, as McIntyre's (1985) study shows. One reason is the occurrence of false alarms, which is a huge problem and really can only be tackled by looking at the situation and not at the alarm or warning. Other reasons for deactivation are likely to follow from the many other problems associated with medical alarms, which we will discuss below. By turning off an alarm, the user is signifying that he or she considers it to be in some way ineffective. A useful programme of research would be to investigate the parameters which are important in determining the turn-off threshold. This would be the threshold at which the negative consequences of an alarm are considered by the operator to outweigh its benefits. It would be interesting to find out which factors are particularly strong in determining this threshold. Such factors could be used to determine alarm design, and for the ergonomic evaluation of existing alarms.

4.2.2.3   Loudness

Without doubt, the loudness of auditory warnings is a crucial factor. If the warning is too loud, it will be irritating and aversive, but if it is too quiet it will go undetected. Loudness is also a very strong factor in determining the perceived urgency of a warning. In Chapter 5 we will be looking at two approaches to the determination of the correct level for auditory warnings. In both of these examples warnings can be produced which will be at an appropriate loudness level for the specific environment in which they are to be used.

In current practice warnings are often so loud that they are turned off before the problem is attended to. The reasons for alarms being too loud in the first place are fairly obvious, as it seems logical when setting the volume of an alarm that it should be set on the loud side to make sure it will be heard in all circumstances. The consequence of this is that the warning will for much of the time be louder than is necessary, which may cause other problems.

Setting the loudness level of a warning requires first determining the background noise level in that environment and then deciding how much louder the warning should be above this background noise. There are other problems which will occur, one of the most obvious of which is masking. The problem with masking in the hospital environment is that the noise level is somewhat unpredictable, as is the potential for two or more alarms to sound at the same time. The problem of spurious masking could ultimately be dealt with by having some appropriate system of central prioritisation of warnings, so that the most urgent and important situation is signalled first, and so on. This may be achieved within the next few years, but it involves compatibility between items of equipment plus additional control technology which has not yet generally been applied to the problem, although relatively simple systems are in use such as the alarm system developed by Schreiber and Schreiber (1989) which will be described later. The question of detection of warnings in noise has been looked at by Stanford et al. (1985) for a set of specially designed warnings which will also be discussed in greater detail later in this chapter. They found that their warnings could be detected against noise up to a signal-to-noise ratio of -24 dB, and even at this level their warnings were detected with 93 per cent accuracy. They predict on the basis of their findings that warnings are likely to be detectable at even lower ratios. For properly designed warnings, therefore, lack of detectability is unlikely to be a major problem. Overly loud warnings are more likely to cause concern. Masking is, however, much more of a problem for more traditional-sounding alarms, as the previously-discussed study by Momtahan et al. (1989) shows.

4.2.2.4 Localisation

Many traditional hospital warnings are hard to localise. McIntyre's (1985) study suggests that anaesthetists do not use the information coming from a warning to localise its direction, but this may be because the warning itself does not assist in such localisation. In other words, the problem may lie in the design of the warning itself and not in the predisposition of the operator. There is little doubt that one of the central purposes of our auditory system is to enumerate and localise sounds in our environment (although this becomes much more difficult in enclosed places), and that it is possible to design sounds which are more or less localisable, depending upon their acoustic characteristics. Unfortunately, the types of warnings typically used in hospitals, such as continuous sinusoidal tones around a frequency of about 1 kHz, are the most difficult sounds to localise. The reasons for this will be dealt with in Chapter 5. Accurate localisation of warning sounds is quite obviously going to improve performance, if only by speeding up the amount of time it takes to walk to the bed from which the warning is coming.

Taking a practical aside for a moment, in discussing the matter of alarm standardisation with practitioners we have sometimes come across the following paradoxical argument. Some people say that they prefer to have a large number of alarms in a unit – and in particular, they prefer to have different patients using pieces of equipment made by different manufacturers, with the different alarm sounds that these inevitably produce – because they claim they use this information to identify which patient's alarm is sounding. In other words, they report that they know which patient requires attention because they know that patient A is associated with ventilator X, and patient B is associated with ventilator Y and so on. Given that this arrangement will be in a continual state of flux due to people coming in and out of the units, at frequent and unpredictable intervals, this is a highly risky strategy to adopt. Here we can see the basis for why they may therefore be wary of a system which uses (for example) a single alarm for all ventilators, a formal proposal which is to be discussed later. Unfortunately, opinions such as the one quoted do not consider the possibility that alarms can be designed so that they are localisable. If they are localisable, one does not need to know who is associated with what – our ears will tell us, directly. This is surely an improvement on the circuitous and risky route outlined above, which is clearly open to failure. The argument in favour of different alarms for each patient also breaks down when we look at the objective data which show that people working in units on a daily basis cannot reliably identify the alarms, let alone make the association between that alarm and the patient to which the piece of equipment is currently attached.

A study by McIntyre and Stanford (1985) analysed the layout of a typical anaesthetic machinery set-up to determine the ability of anaesthetists to identify correctly the source of warnings. They charted the layout of the 'anaesthetic machine', as it is called – a movable table which incorporates gas and vapour delivery systems, ventilator, scavenging equipment, fluid suction systems and monitors. They applied knowledge of spatial hearing to the existing equipment layout and were forced to conclude that, on the basis of spatial layout alone, it would not always be possible to determine which of the alarms was sounding. There are a number of ways this problem could be tackled. The first would be to cut down on the number of alarms, making discrimination easier. The second would be to change the layout of the equipment so that auditory discrimination is easier. The third would be to design auditory warnings so that they are both more easily localisable and more distinct from one another. These questions have been addressed by McIntyre and Stanford (1985) and Patterson et al. (1986).

McIntyre and Stanford (1985) also considered the likelihood of confusability between alarms, and to this end they carried out spectral analyses of the alarms available on the equipment. Detailed acoustic descriptions of the alarms can be seen in Table 4.3. Perhaps the most interesting feature of this set of alarms is that four of them – the airway pressure monitor, the transcutaneous PO_2 monitor, and two of the blood pressure monitor tones – are con-

Table 4.3 List and analysis of a set of operating room alarm signals (from Droh et al., 1985)

	Duration characteristics	Frequency (Hz)	Amplitude characteristics[a]
Airway pressure monitor (Ventilarm 5520, Ventronics)	Continuous tone	2800	Fairly steady amplitude 36 dB
Oxygen monitor (Ohio 200)	Intermittent tone, 660 ms in duration, repeated at regular intervals of approximately 2.5 s	Multiple components: 600, 2250, 2800, 3400, 4000, 4600, 5000	Random amplitude variations 32 dB (average)[a]
ECG (Tektronix 412)	Intermittent tone (analogue with stimulus with patient) 30 ms duration	2000	Low amplitude against some low-amplitude background noise 28 dB[a]
Transcutaneous PO_2 monitor (Kontron Medical 820)	Continuous tone	2900	Random amplitude variation (2–3 peaks per 100 ms) 36 dB (average)
Blood pressure monitor (Dinamap 845, Critikon)			
High tone	Continuous tone	Four components: 900, 1750, 2600, 3400	Steady amplitude 37 dB[a]
Alternating tone	Continuously repeated 240-ms cycle, during which frequency and amplitude of various tones change	Five components: 500, 800, 1200, 1750, 2600	Amplitude of 800-, 1750-, 2600-Hz tones decreases across cycle. Amplitude of 1200-Hz tone decreases from mid-point of next. Amplitude of 500-Hz tone has two high-to-low cycles within main cycle 32 dB average[a]
Low tone	Continuous tone	Four components: 1000, 1650, 2200, 2750	Steady amplitude 34 dB[a]
Minidrip monitor (IVAC 530)	Intermittent tone 400 ms tone 200 ms fade 800 ms silence	3200 with rise to 3400 in last 1/3 of tone	Steady amplitude during tone 38 dB

[a] Amplitude of delivered signal can be changed by operator.

tinuous tones which would be very confusing. In particular, the airway pressure monitor and the transcutaneous PO_2 monitor would be almost impossible to tell apart under any circumstances because they are very close in frequency, they both appear to be single sinusoids and they both have the same intensity level. The two blood pressure tones (high and low) would also be highly confused because they share many characteristics. What an objective, acoustic analysis of the alarms does not show is that it is quite likely that the two high-pitched and the two low-pitched continuous tones would also show high levels of confusion, even though there are vast differences in their frequencies. Of the other alarms, three consist of intermittent tones with quite different cycle lengths. However, we predict that the ECG monitor and the minidrip monitor would be readily confused because there are similarities in their cycle, even though they have quite different spectral characteristics. In fact, the only distinctive alarm is likely to be the blood pressure alternating tone. The reasons why we believe this to be so will become clear in the following pages, and in Chapter 5. McIntyre and Stanford (1985) based their conclusions about confusion by looking at the spectra of the alarm sounds, not by actually applying empirical methods to the question of confusability. However, alarms can look acoustically quite similar, but yet can be quite distinguishable. The reverse is also true. The spectra of many of the warnings shown in Table 4.3 are quite distinctive (with a few exceptions), but this would have little effect on distinguishability if other features such as temporal pattern are similar.

4.2.2.5   Confusion

Confusion is another problem associated with the use of auditory warnings in hospitals and other environments. In addition to the difficulties already covered, which all add to the confusion, there are other problems such as manufacturers using their own 'in-house' warnings on a range of equipment used for different purposes. This may make the manufacturer easy to identify, but it may not improve the user's ability to recognise the function of the equipment. One of the biggest sources of confusion between auditory warnings stems from their shared acoustic properties (Chapter 5). Standards and guidelines on warning implementation often state that warnings need to be discriminable from one another, easily identified and clear, but these statements of what is required do not necessarily tell the designer how to achieve the objectives. There is a relatively large amount of research work on the cognitive psychology of sound which may, however, provide insight into this problem. For example, a piece of research which demonstrates quite clearly one important source of warning confusion by Patterson and Milroy (1980). They looked at the errors made by participants during the learning of a set of fixed-wing civil aircraft warnings and found that warnings which were spectrally quite distinct could nevertheless be confused. The factor which governed if warnings were confused was a feature totally unrelated to spectral quality, and this was the temporal patterning of the warning. Warnings with similar repetition rates were readily confused, regardless of almost any other acoustic feature. Thus the temporal patterning of a warning appears to be a much more salient feature than the spectrum. The acoustic quality of the spectrum is important, however, for other, more physiological and audiological reasons.

The importance of producing warnings which are readily differentiated from one another has featured in the recent attempts to improve and to rationalise hospital warnings. One aspect of their confusion which has been quite prominent in this research is that of the inappropriateness of warnings, which will be dealt in the next section.

4.2.2.6 Inappropriateness

Momtahan and Tansley's paper (1989) makes a number of recommendations about how operating room warnings should be changed, and one aspect that they pick up on is the lack of urgency mapping contained in the set of warnings that they considered. They asked medical-staff participants to rate the urgency of the medical situations that each of the warnings would typically be used to signal. They asked a second group of participants, unfamiliar with medical practice, to rate the urgency of the auditory warnings associated with these situations. This second group of participants was therefore rating only the psychoacoustic urgency of the warning sound. In other words, they were rating only the iconic, not the information part of the warning. They found that there was no correlation between the two measures. Nonurgent medical situations were associated with warnings which were not themselves urgent sounding, whereas some urgent medical situations were associated with warning sounds no more urgent than those used for nonurgent medical situations. Such findings have been replicated (e.g. Finley and Cohen, 1991). Anecdotal evidence also comes from a study of an intensive care unit which showed that a food pump alarm was much more urgent-sounding than a ventilator alarm, even though problems which would cause a food pump alarm to operate would generally be much less urgent than those associated with a ventilator alarm's operating (Meredith and Edworthy, 1994). It may be, of course, that once the information aspects of a warning are known, situation knowledge would override the acoustic qualities of the warning itself – 'Little Bo Peep' would be perceived as urgent if it were known to be associated with cardiac arrest – but as Momtahan's work shows, the meaning of the majority of the warnings used in this particular environment are not accurately recognised. Even if the meaning is known, appropriate urgency mapping could be an additional cue in deciding what sort of action to take, and how quickly it should be taken. Thus greater appropriateness of warnings might be achieved if the relative urgency of the situation could be conveyed by the warning sound, so that the user would know approximately how quickly to respond. Recent studies (Haas and Casali, 1995; Burt et al., 1995) show that simple measures of reaction time correlate with the degree of perceived urgency associated with the acoustic qualities of a warning, but there is much work yet to be done.

4.2.2.7 Summary

Common design practice tends to favour acoustically urgent above nonurgent sounds, for a number of reasons which have been deliberated upon in the previous pages. Generally, auditory warnings used in hospitals sound too urgent. Many of the sounds used are variations of a high-pitched continuous or intermittent tone of the 'How not to design an auditory warning' variety. Some urgent warnings may well be needed because some situations are indeed life-threatening, but many other situations are not. Even when life-threatening problems arise, the alarms used should not be so urgent as to be aversive because this will interfere with the task of the medical staff and will upset the patient yet further. As O'Carroll's (1986) study shows, only a tiny proportion of incidents which lead to the sounding of warnings are actually life-threatening, although many more may become so over a period of time. Proper grading of psychoacoustic urgency, in accordance with medical urgency, might therefore be an important step forward. This philosophy has figured in some of the new approaches to auditory warning design, and the newer types of warnings themselves are more amenable to the modifications necessary to produce such gradations in urgency.

4.2.3   Advances in auditory warning design in hospitals

For auditory warnings to fulfil their potential in the intensive care units and operating theatres, a number of criteria need to be fulfilled. First, they should be designed in such a way as not to be so aversive that the operator wishes to turn them off. Secondly, the number of warnings needs to be reduced, but on a pragmatic, function-driven basis rather than on an *ad hoc* basis. Thirdly, the issue of audibility and masking needs to be addressed. To some extent the number and masking problems interact because if there are fewer warnings they are *ipso facto* less likely to mask one another. Additionally, the warnings themselves should present as few opportunities for confusion as possible. They should be psychologically appropriate in some way and easy to localise. Each of these characteristics will serve to increase the effectiveness of the warnings. The measurement of compliance and effectiveness scores for individual situations, warnings and design parameters remains as a research area for the future. However, there have been several attempts to design improved auditory warnings sets with many of these issues in mind. In addition, there are several standards in preparation which are attempting to elicit these same improvements. These will be considered in the next section.

4.2.4   New designs for hospital auditory warnings

4.2.4.1   *Using vowel sounds*

Many hospital auditory warnings have a sound which is positively aversive. There seems to be an unwritten assumption that for a warning to attract appropriate attention and sound urgent it should be aversive. Perhaps the argument is that if it is aversive the user will want to turn it off by attending to the alarmed situation all the sooner, which is something the alarm designer wishes to ensure. However, what often happens is that the alarm user simply deactivates the alarm. We should note that a warning is a signal, or an icon, representing some situation. The alarm will serve best if it is clearly audible, clearly distinguishable from other sounds and if it sounds appropriately urgent. There is no inherent reason why urgency should be related to aversiveness.

The possibility that users deactivate alarms because they sound too aversive has been addressed by Stanford et al. (1985, 1988). These authors make the logical assumption that a warning system is less likely to be turned off if it sounds pleasing, or at least less aversive. Of course warning sounds do not have to be pleasing, but it is useful if they are not positively unpleasant. To this end, Stanford et al. (1985) designed a set of abstract warning sounds based on vowel sounds. These sounds were not assigned to any particular warning category. They were harmonically rich, with frequency modulation and temporal patterning. In their first study the detectability of the set of eight sounds was determined. The warnings were played at various signal-to-noise ratios against four different types of background noise each likely to be found in an intensive care unit. These were the noise made by a suction device, a skull drill, a grinder-reamer and a plaster saw. The results showed that all of the eight sounds were highly detectable even at a -24 dB signal-to-noise ratio, which the authors say is a far greater ratio than is possible with more traditional, less acoustically rich, sounds. Obviously the more detectable the sounds, the more useful they are in the work environment because they are less likely to be missed, and they do not need to be as loud, which improves the acoustic environment.

In a later study (Stanford et al., 1988) the sounds were tested for affective response. The experimenters measured participants' affective responses to the eight sounds previously designed, together with responses to a set of seven commercially available warnings. The Nowlis Mood Adjective Checklist was used (Nowlis, 1965) which contains 28 mood adjectives (such as defiant, pleased, energetic) clustered into seven more general categories of aggression, anxiety, elation, concentration, fatigue, vigour, and nonchalance. The results showed that the newly designed experimental set of warnings produced fewer affective responses overall than the commercially available warnings. In addition, those responses which were not neutral were more positive to the experimental set than to the commercially-available set. Of 125 non-neutral responses to the commercially available set, 89 were negative. Of 68 non-neutral responses to the experimental set, only 18 were negative. This constitutes a more favourable response to the experimental warnings than to the commercially available warnings. Another finding was that some of the experimental sounds were found to be less acceptable than others in that set.

This study shows that warnings can be constructed in such a way as to arouse fewer negative emotions, more neutral feelings and to be more likeable than is found with traditional warnings. It must be assumed that warnings that do not provoke negative reaction are less likely to be turned off, which in turn is likely to improve their effectiveness. Of primary interest would be the relationship between warning sound parameters, the occurrence of false alarms, workload, and the urge for the operator to turn off alarms, a topic ripe for exploration in research. One point to note is that these warnings are quite unlike most traditional sounds used in hospitals, and so acceptance may be something of a problem. It may be that for a while after the introduction of sounds like these they would fail to be recognised as alarms, which might cause some teething problems. One way to tackle this potential problem may be to set these new alarms initially at a much louder level than that at which they will eventually be needed. Like most new ideas, filtering through to the user takes a great deal of time and for a long period the new ideas may have to sit rather uncomfortably against the old. In some instances this can be a useful aid to discrimination, because receivers might be able to identify alarms by the fact that they are 'new-sounding' in contrast to other alarms that they might hear in the same environment.

4.2.4.2   Rationalising hospital warnings

Another attempt to improve hospital warnings comes from a series of studies, reports and standardisation work emanating from the UK. Initially the problem of too many alarms was highlighted by some UK anaesthetists (Kerr and Hayes, 1983). Their paper highlights the usual problems associated with the use of auditory warnings in the intensive care unit, as well as suggesting that some code of practice is needed, together with a rationale for auditory warning use and design. One of the main focuses of their paper is to reduce the number of alarms while at the same time ensuring that those situations which need to be signalled are accounted for in the alarm system. Kerr (1985) highlights three main criteria which should be met to produce this rationalisation. These are:

1 Each alarm must have the same meaning wherever encountered in the hospital.
2 The number of different alarm sounds must be limited to an absolute maximum of ten.
3 The allocation system must be able to accommodate future technological developments.

Kerr then suggests several ways in which each of these three demands could be met. The first is to have just a single all-purpose alarm, with two or three levels of urgency. This idea has some merit, but limiting the number of auditory warnings to such a small figure does not capitalise on our use of hearing as a natural warning sense. A second way would be one based on priorities, so that alarms would differ in their urgency but would not indicate the nature of the problem, which could then be elaborated by additional information. A third way would involve an equipment-based approach, where particular types of alarm sounds would be allocated to particular types of equipment. The problem with this approach in itself is that it would not prevent the proliferation of alarms. A fourth way would be to adopt a risk-based strategy, which focuses on the point that there are ultimately relatively few causes of tissue damage. Those factors which do cause damage centre around changes in oxygen availability, acid-base status, temperature, drugs, structural integrity of the tissue and radiation. It is proposed that all equipment associated with particular risk categories would have a single alarm which would mean that the range of equipment fitted with a particular alarm could be quite wide. This idea has a great deal of merit because it focuses on the patient, rather than the equipment, which of course is the thing most resistant to change, thus fulfilling the third (and by luck, the second) of the requirements highlighted above. Kerr (1985) therefore proposes a 'risk and response-based' strategy, which is a combination of the equipment-based and the risk-based approach. He proposes the use of six alarms, allocated according to both risk to the patient and the response required by the attendant. For each of the six categories Kerr also proposes the use of a cautionary and an emergency version. The categories are as follows: hypoxia, ventilatory problems, cardiovascular problems, interruption to artificial perfusion, drug administration and error, and thermal risk. Kerr shows how a wide variety of problems can be allocated to each of the six categories, which are shown in Table 4.4. This lists a wide variety of medical situations in which an alarm might be needed, showing whether the cautionary or the emergency version of the alarm should be used in each individual case. The table also shows which of the six warning sounds should be used for each particular medical situation.

This philosophy is being carried forward into the British Standard for alarm devices in intensive care units and operating theatres currently being developed by Health Care Committee 16 of the British Standards Institute (the committee is called 'Medical alarms and signals'). A set of auditory warnings based on the principles advocated by Patterson (1982) has been designed and is proposed in the standard. A central theme of the risk-response based system is that each situation should have a particular alarm sound. Specimen alarm sounds are proposed for each situation (Patterson et al., 1986). Less restricting suggestions are proposed in draft worldwide and European standards (EN 475 and ISO 9703-2) which will be reviewed later.

The system of specifying alarms according to risk and response-based categories would immediately solve the problem of proliferation. Warnings designed according to Patterson's guidelines would also result in warnings that are more readily localisable (as are Stanford et al.'s (1988) proposed sounds) and, at least potentially, more distinguishable. They do not appear immediately to address the question of localisation, but the way in which the spectra of the warnings are specified (requiring them each to possess at least four harmonics between the 500–4000 Hz range) means that they will be much easier to localise. The forthcoming standards also have something to say about

Table 4.4 Allocation of alarm sounds according to a risk-and-response based system. In the risk part of the table, the type of sound is indicated by a capital letter (H = hypoxia; V = ventilation problem; C = cardiovascular problem; P = artificial perfusion device problem; D = drug administration problem; T = thermal risk) and the form by a small letter (e = emergency, c = caution) (from Kerr, 1985)

Measured variable	(a) Priority level		(b) Risk
	Emergency sound	Caution sound	Type of sound
Brain oxygen (low)	+		H e
Intravascular oxygen (low)	+		H e
(high)		+	H c
Trans cutaneous oxygen (low)	+		H e
(high)		+	H c
Gaseous oxygen (low)	+		H e
(high)		+	H c
Oxygen supply (if air NOT substituted)	+		H e
Oxygen supply (if air substituted)		+	H c
Minute volume (low)	+		V e
(high)		+	V c
Airway pressure (low)	+		V e
(high)	+		V e
Apnoea indicator	+		V e
End-tidal carbon dioxide		+	V c
Cardiovascular monitors	+		C e
Balloon assist devices	+		C e
Haemodialysis			
Air	+		P e
Disconnection	+		P e
Low flow		+	P c
Ultra-filtration rate		+	P e
Fluid temperature (high) (1)	+ ?	+ ?	T ?c
Plasmapheresis			
Air	+		P e
Disconnection	+		P e
Low flow		+	P c
Fluid temperature (high) (1)	+ ?	+ ?	T ?c
Syringe pumps (2)	+ and	+	D e and c
Infusion pumps (2)	+ and	+	D e and c
Infusion controllers (2)	+ and	+	D e and c
Inhalation anaesthetic meters	+		D e
Enteral feeding pumps (3)			
Incubator temperature (high) (1)		+	T c
(low)		+ ?	[T ?]
Radiant heater temp. (high) (1)		+	T c
(low)		+ ?	[T ?]
Humidifier temp. (high) (1)		+	T c
Heating blankets (high) (1)		+	T c
Brain activity (EEG)	+		H e

appropriate sound levels. The forthcoming European standard (EN 475) gives a generous range of amplitude levels, ranging from 45 to 85 dB depending upon ambient noise levels, measured according to ISO 3744, 'Acoustics – sound power levels of noise sources: engineering methods for free-field conditions over a reflecting plane'. As EN 475 is meant as a generic standard, on top of which may be built specific equipment standards, it is quite likely (and the standard draws attention to this point) that the range will be further restricted when the particular environment in which the device is to be used is taken into account.

The philosophy of warning prioritisation, but without precise specification of a single warning sound, is being drafted by the European group under Technical Committee 259 in the form of EN475. In this proposal, the idea of having six specific warning sounds is not being considered. In this standard, three general warnings are proposed; a high, medium, and a low priority sound. Each of these sounds differs in their level of urgency, and rather than having precise warnings specified in the standard, quite generous degrees of freedom are specified. Within these degrees of freedom many versions of each of the sounds are possible, although any resulting warning will remain recognisable. This gives the manufacturer some scope in specifying the sound, but also ensures that the warnings come recognisably from a 'prototype' sound, which is embodied in the standard itself. This approach capitalises upon the natural predisposition of the cognitive system to group and to categorise sounds, and is one that could be exploited to potentially great effect in auditory warning design.

4.2.4.3   Simple prioritised systems

Another approach to alarm rationalisation is that developed by Schreiber and Schreiber (1989). They have developed a structured alarm system which considers the time taken for the human observer to react to the situation, and the increasing potential for escalating medical problems as time passes. In this approach, the whole process from initial sounding of an alarm to a potentially critical situation is considered in a type of closed loop with feedback. Incorporated into this idea is the use of three levels of urgency of alarm. These consist of a warning, which requires immediate response, a caution sound, requiring prompt response, and an advisory signal, which requires only awareness. An interesting aside here is that the words 'Warning' and 'Caution' are used to designate different levels of urgency. Several studies of signal words have shown that people's ratings of these words are very similar and that they are generally associated with approximately the same degree of risk, which is something of a problem with respect to the calibration issue. This has been discussed at length in Chapter 2. Returning to the auditory warning issue, the idea behind Schreiber and Schreiber's implementation is similar to that proposed by Kerr (1985) and is one of the central ideas incorporated into the new standards, as discussed earlier. These three categories aim to elicit the appropriate response from the hearer of the alarm. Whether they do or not is another topic for future research. In indicating levels of urgency, different types of sound for each of the three categories are advocated. For the warning sound deemed to be the most urgent, a continuously repeating sound pattern is used. For a caution sound, an intermittently repeating sound is used and for the advisory category a single short tone, or no sound, is used. Together with this proposition goes the idea that only the alarm sound corresponding to the most urgent situation should be allowed to sound, and that all other alarms should be temporarily suppressed. Once the most urgent situation has been adequately signalled, the next highest priority is signalled and so on. Thus Schreiber and Schreiber (1989) address the issue of the centralisation of alarms, which is usually absent from the standards which tend to deal rather more with the sounds themselves, and the specific equipment to which the standard is directed.

4.2.5   Overview and practical problems

As the preceding pages show, there are a number of approaches that have been taken in tackling the numerous problems associated with hospital alarms, especially those used in the operating theatre and the intensive care unit. Some of these approaches have looked at the philosophy behind alarm systems (such as Kerr, 1985) while others have concentrated on the acceptability of alarm sounds (such as Stanford et al., 1985, 1988). Others (Schreiber and Schreiber, 1989) have looked at the 'alarm situation' and focused on the time progression of potentially dangerous situations, and on the centralisation of alarms. Each of these approaches is likely to improve the effectiveness of alarm sounds inasmuch as numbers could be reduced, warnings could be less confusing, more easily localisable, more appropriate in matching the urgency of the situation being signalled, and alarms could be prioritised in some way. None of them, however, directly addresses the question of effectiveness, although increased effectiveness is clearly implied by such developments. In practice it seems that strong resistance is encountered to any alarm sounds which sound new, or different. It seems that people are very concerned about the aesthetics of alarm sounds. For example, one feature of the response encountered when introducing non-traditional alarm sounds is whether or not they are liked by the people who will using them. This does not appear to have been an issue with traditional alarms, but it seems to present a focus for people who do not wish to have the alarm system changed. The only requirement of warning sounds should be that they are not positively disliked, because it is dislike of an alarm system which causes people to turn it off. After all, it is not likely that people will really want to listen to alarm sounds in their leisure time.

Why these reactions to new sounds occur may be of psychological interest, but that is another story. A further practical problem in trying to get new warning sounds into use is that reactions to standardisation are often political and self-protecting rather than rational and scientific. The most radical and complete solution to the problems with alarms in intensive care and operating theatres is to standardise on a warnings philosophy through the official standards bodies, and to place some constraints on the types of warnings that can be used. The most extreme way of doing this is to specify warning sounds precisely in the standard itself. Another method is to allow manufacturers some degrees of freedom in the specification of the warning sounds in the standard, but to write the standard in such a way as to allow each sound always to be recognisable as a particular sound, no matter how it is constructed within the parameter limits laid out. A third approach might be to specify only the spectral content of the warnings, and allow manufacturers to develop their own systems otherwise. This will not stop proliferation and confusion, but might reduce the problems of localisation and audibility, which in itself would be a step forward even without further refinement.

4.3   Helicopters

4.3.1   The Sea King project

The Sea King helicopter used by the UK military establishment has traditionally used few if any auditory warnings, relying heavily on visual warnings. However, besides the potential for missed visual warnings, pilots today carry out a larger proportion of night flying which requires the use of night vision goggles. These goggles improve vision outside of the helicopter, but restrict vision rather alarmingly within the helicopter. Auditory warnings are therefore essential under these circumstances, but they should not be allowed to proliferate. Fortunately, the visual warning system already in existence is relatively simple. Three categories of response type are indicated by three coloured lights. These are red for immediate action, amber for immediate awareness (where no immediate action is needed) and green for information available (where no action is necessary). Thus a system of colour coding – indeed of urgency mapping – is already in operation. The primary task in the design process was to generate alarms which matched these priorities. An auditory warning system based on the same philosophy was proposed (Rood et al., 1985) and a set of auditory warnings was developed on the basis of this proposal (Lower et al., 1986).

The warnings were constructed according to guidelines developed by Patterson (1982). These guidelines will be described in detail in Chapter 5, as will the detailed procedures involved in determining appropriate loudness levels and the 'designing' of perceived urgency into the warning sounds themselves. One particular warning is described here, a priority-one warning for 'Fire'. The basic building block of the warning was a pulse of sound lasting 150 ms, with a fast onset and slow offset envelope. The fundamental frequency of the pulse was 240 Hz, and the pulses possessed 7 partials. These partials were alternately shifted by 10 per cent above or below the integer multiple of that particular partial for distinctiveness. Finally, each of the pulses was weighted appropriately for loudness. Because the noise in the helicopter is higher for lower frequencies, the lowest partial at 985 Hz was made more intense than the highest partial, which had a frequency of 2425 Hz. The pulse was used to build up a burst of sound, which is shown in Figure 4.2.

This shows that the burst consisted of 6 pulses, the first two of which were played at a lower pitch than the last four (the P values shown in the figure). The figure also shows that the attenuation of the pulses varied relative to one another (the A values in the figure), with the first two pulses being at a lower level than the rest, and with a slight reduction on pulse 5. The pulse spacing values (PS) are shown underneath these two sets of values, showing that the burst had a regular rhythm. Three versions of the burst are shown: an initial burst, set at a moderate level of urgency, an urgent burst, set at a higher level of urgency, and a background burst, set at a lower level of urgency. Different versions of the burst were used in a complete warning, which consisted of several playings of the burst. The precise order of initial, urgent and background bursts depended upon the priority of the situation. Figure 4.3 shows examples of how a priority-one and a priority-two warning could cycle. In the warning described here, for example, the initial warning (I) would be heard first, followed by a voice message (VM) (for redundancy). This would be followed shortly by the urgent burst and a voice message. The warning would

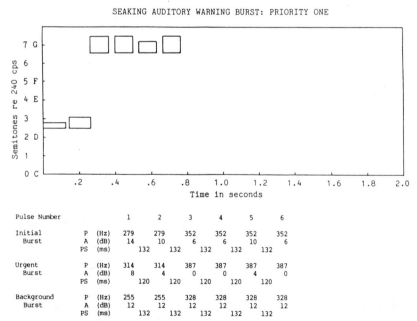

Figure 4.2 An example of an advanced helicopter warning. The warning consists of 6 short pulses. The pitch values (P), attenuation values (A) and pulse spacing values (PS) for each individual pulse is shown. Three versions (initial, urgent and cautionary) versions are shown (from Lower et al., 1986).

PRIORITY ONE WARNING

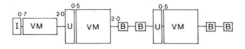

PRIORITY TWO WARNING

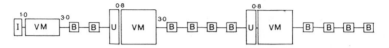

Figure 4.3 Structure of complete warning (Priority 1 and Priority 2). Each box represents either a warning burst (I = initial, U = urgent, B = background) or a voice message (VM). Times between parts of the warning are shown (from Lower et al., 1986).

then cycle twice in a background form (B), to allow communication above it, but then if the situation was not alleviated the urgent burst (U) would return and so on until the situation was rectified. In priority-two warnings, shown underneath, there would typically be more cycles of the background form before the warning returned in its most urgent, attention-grabbing version. The background version of the burst would be set at such a level as to allow speech to be heard above it quite easily. Note here that attenuation levels are stated rather than actual loudness levels, which means that the individual pulses would be reduced by that number of decibels relative to a specific intensity level. The actual intensity level at which the warnings would need to be presented is to be described in the next section. Attenuation levels are relative to the loudest pulse in the loudest burst (pulses 3, 4, and 6 in the urgent form of the particular warning), which would not be attenuated at all relative to the specified standard.

4.3.1.1 Establishing loudness levels

The first phase of this project was the establishment of appropriate loudness levels for auditory warnings components. First, narrowband noise readings were made in several different flying conditions and a range of noise spectra was produced. These spectra show loudness levels for each of the bands measured. The next phase of the project was to predict auditory threshold across this range. This is not a straightforward task, but depends upon the way in which the auditory filter functions for different centre frequencies. The auditory filter is a theoretical concept which models the functioning of the auditory system in its ability to detect tones. If two tones are close in frequency and are of approximately the same level of loudness, it is likely that one will mask the other. If, however, two tones similar in loudness but further apart in frequency are heard together, masking is less likely to occur. Thus the auditory system can be thought of as a bank of filters with specific attenuation characteristics.

As the attenuation characteristics vary with centre frequency, the relationship between the noise spectrum and threshold varies across the noise spectrum. Threshold cannot be assumed from the noise spectrum, but has to be calculated. This is, of course, most easily done by computer. A suite of programs housed at the Institute of Sound and Vibration Research at the University of Southampton was used to carry out these calculations. The actual noise spectrum derived for the Sea King helicopter is shown in Figure 4.4. Figure 4.5 shows the threshold curve for this noise environment.

Once auditory threshold in the worst case spectrum had been established, the appropriate level for auditory warnings was set at 15–25 dB above this. These lines are shown in Figure 4.6. The important thing here is that the appropriate range varies with frequency – a low frequency component of a warning will need to be set at a much higher level than a higher frequency component, because the noise spectrum, and hence threshold, is lower at higher frequencies.

The final phase in establishing levels was to take account of the frequency response of the headphones through which the warnings were to be heard, and to predict the voltage levels required to produce the appropriate levels for the warning components established from the previous phase (Lower and Wheeler, 1985; Lower et al., 1986). At this point, the relative levels for the components of any warning have been clearly established.

4.3.1.2   Warning design

The next phase in the project was to specify and design the actual warnings. After several meetings and much discussion it was established that there were

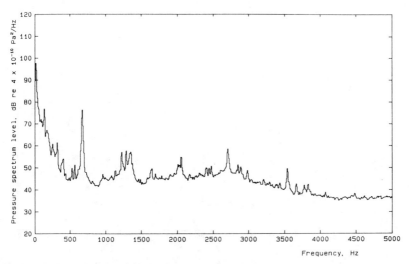

Figure 4.4   Noise spectrum of a Sea King Helicopter measured at the ear (from Lower et al., 1986).

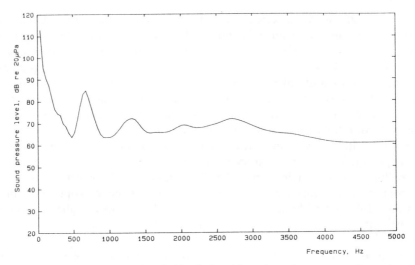

Figure 4.5   Masked pure-tone threshold calculated from the noise spectrum shown in Figure 4.4 (from Lower *et al.*, 1986).

five or six clearly separable problems which could result in the need for a priority-one, immediate action response – those signalled by red in the visual warning system. There were also several situations in each of the priority 2 (immediate awareness) and priority 3 (information) categories during flight. Proliferation, or at least the potential for it, had therefore reared its ugly head.

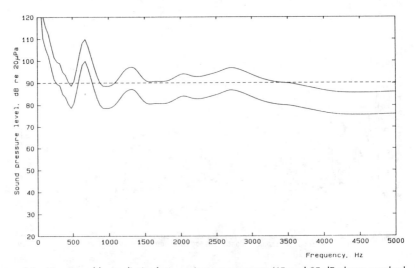

Figure 4.6   Upper and lower limits for warning components (15 and 25 dB above masked threshold as shown in Figure 4.5) (from Lower *et al.*, 1986).

The strategy developed to cope with this problem was to assign different warnings to each of the priority-one situations. All of these situations would require a response in 2 or 3 seconds, so neither a visual nor a spoken warning would be particularly viable. However, for categories other than the top priority there would usually be enough time to search or wait for other information, which should help clarify the situation. For this reason it was decided to use a single warning to annunciate each of the lower priorities, and then to provide additional information concerning the problem using a spoken message and a visual warning or both. In other words, there is time in this situation to present both the icon, and the information describing the actual situation which the icon represents. Thus in the amber and green conditions the warning would signal only the priority of the warning and not the nature of the problem itself, unlike the priority-one warnings. This approach kept the total number of warnings required down to eight, which falls within a learnable range. A set of warnings corresponding to these priorities was designed according to Patterson's principles (Patterson, 1982). Two additional special warnings were added, one a 'No track' warning and the other a 'Low height' monitor, making a total of 10 warnings. According to the state of the art of perceived urgency at that time – these warnings were designed some years ago – the priority-one warnings were designed to be more urgent than the priority-two warning, which in turn was designed to be more urgent than the priority-three warning. An example of one of these warnings was given in Figure 4.2.

4.3.1.3 Refining the warnings

At this point there was a potential, and indeed a necessity, for further design refinements. Obviously the six top priority warnings were intended to be as discriminable from one another as possible, but they were all expected to be of a relatively high level of perceived urgency. In design terms, this is a somewhat taxing problem because warnings which are similar in urgency may sound more similar to each other than would be ideal for reliable discrimination between them. In addition, the urgency mapping of the warning set was important. Urgency mapping was achieved in the first design phase by manipulating the more obvious features of a warning such as its pitch, its temporal pattern, its attenuation characteristic and so on. The 'first draft' of these warnings was tested for confusion and canvassed for perceived urgency (Chillery and Collister, 1986). They found that confusions between some of the warnings in the set occurred a little more often than would be ideal, and in addition some of the top priority warnings were thought by the pilots who would eventually be using them to be not urgent enough. In response to this finding a revised set of 10 warnings was prepared for testing, using five of the original warnings and five new ones designed with these joint problems in mind.

The new warnings were tested for confusion on a population of pilots as well as a population of civilian defence personnel, using a self-paced cumulative learning paradigm. Participants were played a single warning, told to repeat its name, after which they were then played a pair of warnings. This pair would include a new warning, plus the one already learnt. On naming both correctly, the third warning would be introduced, and so on. The experiment was carried out with participants seated inside a Sea King helicopter, listening through the headphones in simulated noise so as to copy as far as possible the acoustic situation in which the warnings would typically be heard (Chillery and Collister, 1988). The flying task itself was not simulated, however. The results of the experiments showed that a small number of confusions still existed but that the number had fallen in relation to the original set of warnings. The results also indicated that civilian personnel responded in very much the same way as pilots, except that generally their performance was a little lower, which is useful to know because it enables testing to be carried out more quickly and cheaply. Another feature of the results was that they showed that the set of pilots tested were able, after retesting, to learn and recognise a set of 10 warnings with 95 per cent accuracy.

A further test was then carried out on the perceived urgency of the set of warnings. A larger set of other warnings was prepared and tested for these trials. The additional warnings were intended as additional priority-one warnings, with the lower priority warnings remaining the same as before. Participants performed a paired comparisons procedure whereby each of the warnings was paired with each of the others, and rankings data obtained. From this experiment the six most urgent warnings were selected. These were then added to the other four original warnings and tested for confusion as before (Chillery and Collister, 1988). Only a very small number of confusions was identified, thus the warning set was deemed acceptable in design terms.

4.3.1.4 Testing the warnings

At this point a set of 10 warnings that were largely resistant to confusion and appropriately urgency-mapped had been obtained through an iterative design and testing process. The next stage was to determine whether any advantage might be attached to using warnings of this type over and above more simple, possibly less costly, warnings. The new warnings were less irritating and aversive (which is in itself a useful step forward in ergonomic terms) and it was important to demonstrate that there might also be performance advantages associated with these new warnings. To this end a study was carried out on the relative merits of three warning formats (James and James, 1989). These three formats were those that might typically be found in situations where auditory warnings are to be used. In one condition a voice message alone was used, a simple tone followed by a voice message was used in the second condition, and in the third a prioritised warning of the type specifically designed in this project was used. This study was carried out in a quiet laboratory rather than in a simulator. Participants were given a keyboard, with each of the situations associated with the warnings represented by a key, which was to be pressed during the experiment when the appropriate warning was heard.

Before carrying out the experiment each subject was trained in the meanings of the warnings. The experiment was a within-subjects design, so all participants sat through six randomised sessions where they were required to respond to warnings in the three types of auditory format, using two types of keyboard format (where the keyboard was either completely randomised, or blocked according to the three priorities discussed above). The results of this experiment showed that there was a significant effect for warning format. The no warning (voice only) condition produced the slowest response times while the prioritised warning format produced the fastest response times. The blocked keyboard also produced faster reaction times than the scrambled keyboard. There were other fairly complex effects and interactions but the main result was a vindication of the design process. The prioritised warnings produced a faster response time than either the simple tones or the voice-only messages.

The extent to which response time is a valid reflection of the effectiveness of auditory warnings for helicopters must be considered briefly. It is clear that, as a first test of effectiveness, a response time paradigm is both useful and convenient. Some of the responses also required under the conditions in which such emergencies occur are simple physical responses (such as bailing out). Thus, response time is, at least to some extent, both a valid first dependent variable in the measurement of effectiveness, and for helicopters in particular it can resemble the type of response actually needed in some real flying emergencies. Faster reaction times obtained from experiments where participants are explicitly told to respond quickly in either a simple or a choice reaction time experiment can also be seen as indicating the superiority of this type of warning.

The ultimate effectiveness of the new warning system designed in this collaborative project, as with most other warning design projects, is rather harder to gauge. Warnings are heard only rarely when flying. As Rood points out (1989), it is hoped that a pilot will never see or hear a warning in his or her flying career. Thus simulating anything like normal warning occurrence would prove to be costly both in flight testing and in simulators. In the meantime, a commercial auditory warning system programmed with the warnings just described has been flown in experimental Sea King and Lynx helicopters. Some importance has also been attached to standardisation, with the possibility of many different types of military helicopter being equipped with the same warnings, so that they are recognisable by pilots when they move from vehicle to vehicle. To this end it has been proposed that this warning set goes forward into the new EH101 helicopter, and it is also proposed ultimately to standardise auditory warnings across all military aircraft. In the UK a minimal set of warnings is currently flying in the Lynx helicopter. It consists of two each of priority-one and priority-two warnings, one priority-three warning, a 150-feet warning and a low height warning, making a total of seven.

It is generally expected that the inclusion of properly designed warning sets, where consideration has been given to the acoustic environment, the prioritisation of alarms, the design of the warning sounds themselves where confusion and perceived urgency have been duly considered, and the testing of the warnings using relevant dependent variables, will enhance and extend the flying envelope of such helicopters. However, the ultimate question of effectiveness is very hard to gauge because of the cost of testing and the low frequency of alarm use. The issue is further complicated by the cost/benefit decision as to whether to comply with the warning, which can only really be considered when some of the more fundamental issues have been cleared up. Despite the current lack of an ultimate test of the effectiveness of this warning system, all the indicators are that helicopters will be well served by such a system, and that the risky business of flying a helicopter will be made safer.

The project as a whole demonstrates how each of the important characteristics of warning design and implementation can be addressed in collaborative work. This work does, of course, cost money. In helicopter flying the cost of building helicopters and training pilots runs into many millions, and so the costs of developing warning systems can be absorbed by such costs and recouped easily. If one warning helps to keep one helicopter up in the air, a helicopter and a pilot may be saved and the entire warnings project paid for many times over. Aside from the human cost, the monetary benefits are also obvious.

CHAPTER FIVE

Nonverbal auditory warnings

5.1 Detectability

5.1.1 Introduction

One of the central problems in auditory warning design and evaluation is that of detectability. Because of the temporal and ephemeral nature of sound, calibrating auditory warnings correctly in terms of their loudness presents the ergonomist with a whole new set of complex problems that are not relevant to visual warning design to quite the same extent. Perhaps the closest parallel here is that of the placement of a warning label. Research shows (Frantz and Roades, 1993) that the placement of a warning label can have a significant impact on levels of warning compliance, particularly if it breaks up the flow of a task, forcing the person using the equipment to at least look at the warning. Since our ears are always open to any auditory stimulus regardless of where we are looking, or what we are doing, auditory warnings are able to break up the flow of our actions without much effort; it is in fact their natural tendency. However, it is not sensible for auditory warnings to break up the flow of activity too much, because they tend to be used in situations where the operator is busily occupied with other important ongoing tasks, rather than in the typical warning label situation where a user may be using a potentially dangerous product and needs to be informed of the likely risks.

Current practice is such that in generally quiet environments auditory warnings are rarely too quiet but are often too loud. In noisy environments, the potential for missed warnings greatly increases, although in some noisy environments warnings are still sometimes too loud. An additional problem may be that the overall level of noise in the work environment varies quite significantly over time, which is yet another problem for the designer to overcome.

For each of these three types of environment – the generally quiet, the noisy but relatively fixed, and the fluctuating – the determination of auditory threshold is of central importance because the appropriate level for auditory warnings will always be some specifiable and fairly consistent number of decibels above this threshold. In a relatively quiet environment, the appropriate level for an auditory warning is likely to be considerably above threshold because of the need for them to be attention-getting and to communicate some sense of urgency. In noisy environments the appropriate level for auditory warnings will be closer to threshold because, although absolute levels do not determine audibility (the audibility of a sound is usually determined by the level and frequency of noise in which it is heard) very high absolute levels can be damaging to hearing and can lead to inappropriate behaviour when warnings are heard. Such behaviour includes the turning off of warnings, startle reactions and so on (Thorning and Ablett, 1985: Rood *et al.*, 1985).

In a fluctuating noise environment the most efficient approach may be to establish which warnings and other significant sounds in that environment are likely to occur together, and then to determine which of those are likely to mask each other if heard together.

5.1.2   Predicting auditory threshold

The absolute threshold for hearing is, of course, extremely low and is usually of little practical use in most acoustic environments. It has no relevance in situations where auditory warnings might typically be used where people are working or talking, where pieces of machinery are in operation and where the noise levels may be quite high for some reason. It is more elucidating to think of working environments as being fixed or variable in their noise levels. Determining threshold and the appropriate level for auditory warnings in a fixed noise environment is easier than in a fluctuating noisy environment, although some suggestions will be made later on in this chapter as to how some steps may be taken to minimise the impact that unpredictability will have on auditory warning detectability. The most pragmatic solution here may be not to develop some complex acoustic algorithm, but to rationalise the warning system in other ways.

Aircraft are good examples of environments where the noise level is relatively fixed. The noise spectrum of the cabin of a fixed-wing aircraft is relatively constant, although it can increase in level as speed increases and as different manoeuvres are undertaken. However, recommendations about changes in level can be made on the basis of spectral information obtained under different flying conditions. One approach to the problem of determining auditory threshold for a fixed noise spectrum is to determine the auditory threshold empirically across the whole noise spectrum, although this is a very time-consuming and expensive procedure. A way of circumventing this is to model the functioning of the ear, based on noise measurements from the environment under consideration, and to predict threshold from this model-ling procedure. Patterson (1982) demonstrates one method of doing this which has been successfully applied in a number of warning design projects (Lower *et al.*, 1986; Edworthy and Hellier, 1992a). Today, several versions of the soft-ware which allows the prediction of masked threshold are used in acoustics laboratories around the world.

The central problem in predicting threshold for a sine tone or a more complex sound in noise is that the detectability of a sound is not just deter-mined by its intensity level, but by its frequency and the amount of noise also present around that frequency value. Thus, as Patterson (1982) points out, a noise can have a million times the acoustic power of a signal, and yet the signal can remain perfectly audible if the bulk of the signal power occurs in a different frequency range than the spectrum of the noise. On the other hand, a noise will mask a signal if the power spectrum of the noise is greater than the signal at every point in the spectrum. This is the fundamental logic that has also determined the development of a slightly different model of threshold prediction, described by Laroche *et al.* (1991), which will be described later in this section.

When a listener hears a tone or sound he or she makes use of a selective filter which focuses at the frequency of the tone, filtering out noise as best it can. Like any physical filter, this auditory filter has a measurable bandwidth; as the power spectrum of the noise and the sound come closer together, so more noise gets in through the filter until the signal is eventually masked by the noise. The modelling of the auditory filter is central to both methods for predicting audibility which will be presented here. They are essentially both 'power-spectrum' models.

5.1.2.1 Patterson's approach

The auditory system can be likened to a bank of such filters, and so the detect-ability of a signal in noise depends critically on their operation, each filter focusing different frequencies or groups of frequencies. Thus the shape of the auditory filter in terms of frequency response and attenuation characteristics is all important. Patterson (Patterson 1974; 1976; Patterson and Nimmo-Smith, 1980) measured and refined the concept of the auditory filter in a series of experiments by using the standard 2-alternative forced-choice paradigm. He determined threshold at various points in the spectrum, in noise and in quiet listening conditions, coming to the conclusion that the bandwidth of the audi-tory filter decreases from about 14 to 11 per cent of the centre frequency (the frequency upon which the filter is focused) as it increases from 500 Hz to about 4 kHz. As a rule of thumb, the bandwidth of the auditory filter can be thought of as being approximately 15 per cent of the centre frequency. The filter is also approximately symmetrical, provided that frequency is plotted in a linear, rather than a logarithmic, scale; and its attenuation characteristic drops by approximately 100 dB per octave.

Patterson's (1982) guidelines for the construction of auditory warning systems use this auditory filter model as the basis for predicting thresholds. Figure 5.1 shows the noise spectrum of a fixed wing aircraft, together with threshold calculations based on Patterson's model of the auditory filter. The lower solid line shows the spectrum of level-flight noise, and the one above it shows the auditory threshold for this environment. One can see that the noise spectrum line and the masked threshold line are not exactly parallel. In general, as frequency rises so the difference between the two lines increases. Thus the determination of threshold is not simply a question of drawing a single line a fixed amount above the noise level at each frequency.

Once threshold has been established, the appropriate level for auditory warnings and their components can be proposed. A variable noise spectrum (variable over frequency, not time) will produce a variable threshold line; however, all points along that line are in some sense equal because they are the level at which detection of a tone within the noise in question can be

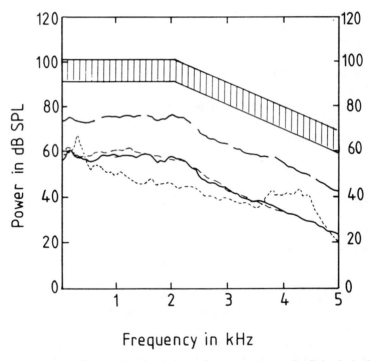

Frequency in kHz

Figure 5.1 The range of appropriate levels for auditory warnings on the flight-deck of a Boeing 727. The lower solid line shows the spectrum of level-flight-noise. The broken line above this shows auditory threshold, and the two lines above this are 15 and 25 dB above threshold, showing the appropriate region for auditory warning components (the shaded region). The faint and dotted lines show the spectra of the flight-deck noise during steady climb and steep descent (from Patterson, 1982).

achieved 75 per cent of the time in a 2-alternative forced-choice paradigm (this is the convention used by acousticians). As intensity increases above threshold so detectability increases to an asymptote which is such that by 15 dB above threshold signals are difficult to miss. Patterson suggests, therefore, that for auditory warning signals the lowest level at which they should be presented is 15 dB above threshold. He also proposes an upper limit of 25 dB above threshold, on the basis that as sounds rise above 90 dB(A), which might be necessary in some aircraft as well as in other work situations, they become aversive very rapidly. In Figure 5.1, the appropriate band for warning components within that environment is also shown. Increasing the loudness of a stimulus makes little impact on its detectability when the loudness is more than 15–20 dB above masked threshold, but can increase its adverse characteristics. This is an important point in so-called quiet environments, where alarms are often much louder than necessary.

One important aspect of warning design, which will be considered later, is that several harmonics should be present, for two reasons – to reduce the risk of masking and to enhance the localisability of the warning. For this strategy to be effective each of the components of the warning must lie within Patterson's 15–25 dB band. For example, a warning with a 200 Hz fundamental and a set of related harmonics at 400, 600, 800, 1000, 1200, 1400 and 1600 Hz should have each of the harmonics tailored to the appropriate level for that component. Figure 5.1 shows that the appropriate band for warning components (the shaded area) is lower for higher frequencies. Thus the higher frequency components should be considerably less intense than the lower ones. Such tailoring was carried out in the helicopter project described in Chapter 4, where the noise spectrum, and thus the masked threshold, is complex because of large amounts of noise at low frequencies, in addition to noise peaks in the mid- and high-frequency range caused by the gears.

Many warnings currently in use consist of only one or two components. These components are in practice often masked, so the warnings tend to be set at excessively loud levels. Another problem can be that warnings are inappropriately loud or quiet in the wrong parts of the spectrum, given the noise environment. Figure 5.2 shows the same masked threshold and appropriate band for warning components as shown in Figure 5.1, but with the components of the fire bell used on the flight-deck of a BAC 1–11 superimposed. This shows that the fire bell warning has 10 components, eight of which are below threshold. One of these is actually within the appropriate band, and one is too loud by about 15 dB. Overall, this warning will give the impression of being louder than is necessary, even though only two of the components are actually audible.

5.1.2.2 The 'Detectsound' model

A slightly different method of predicting auditory threshold comes from a group based at the University of Montreal (Laroche *et al.*, 1991) who have

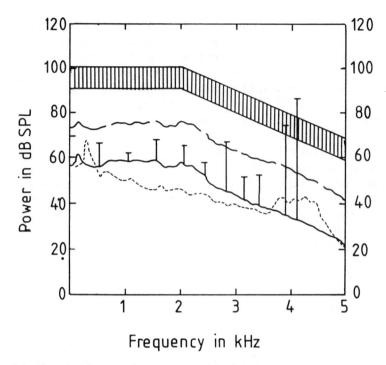

Figure 5.2 The principle spectral components of the firebell (solid vertical lines) on the flight-deck of a Boeing 727. The components are superimposed on Figure 5.1 (from Patterson, 1982).

developed a set of computer programs known as 'Detectsound'. This set of programs models the functioning of the ear in detecting warning signals. It differs from Patterson's method because it uses a slightly different model of the auditory filter (Zwicker and Scharf, 1965). Zwicker and Scharf also conceptualise the ear as operating like a bank of filters, but use different methods to calculate the critical bandwidth, which ultimately determines if masking will occur. Laroche *et al.* (1991) point out that Zwicker and Scharf's model has been validated and adopted as a standard. Frequencies covered by that model's critical bandwidth calculations run from 63 to 12.5 kHz, which is a larger range than Patterson's, and the sound pressure levels for which the model claims to be valid extend to 100 dB, which is higher than, for example, the limit claimed by Patterson and Moore (1986). In noisy workplaces it is important that a large range of frequencies and intensities is covered.

The Detectsound model is also able to consider the effects of age on both auditory sensitivity and frequency selectivity, which allows the model to function in a more generalisable way. Work by Patterson shows how the critical bandwidth, the portion of the noise spectrum which actually masks a tone at a specific frequency, increases by about 2 per cent for each period of 10 years after the age of 20. The calculation of the appropriate bandwidth given the age of the hearer is thus incorporated into the Detectsound model. The aim of the model is to provide an excitation pattern for the noise environment, as well as for each warning heard in a specific environment. These excitation patterns model the activity of the ear when the noise, and the warnings, are heard. Comparing these excitation patterns shows the potential for masking. Here, one has to consider not only the potential for noise to mask warnings, but for warnings to mask one another. Unlike Patterson's model, which uses precise Fourier analysis to arrive at its conclusions, Laroche *et al.*'s (1991) model uses $\frac{1}{3}$ octave band measurements. These have the disadvantage of being less precise, but the advantage that they are cheaper to acquire and more convenient to process.

The Detectsound model involves several stages of calculation, shown in Figure 5.3. The first is the calculation of a correction factor for the use of hearing protectors, if necessary. Often workers in noisy workplaces will be wearing hearing protectors which will need to be considered. The effect of wearing hearing protectors on the perception of both speech and warning

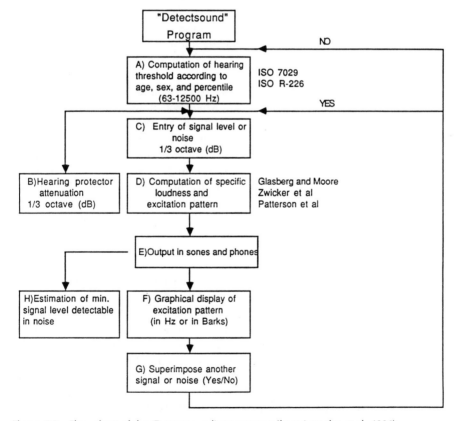

Figure 5.3 Flow chart of the 'Detectsound' programme (from Laroche *et al.*, 1991).

sounds is a complex issue and beyond the scope of this book. Wilkins and Martin (1987) provide a useful review of this issue. The second part of the calculation in the model involves corrections for the transmission factor. The transmission factor is a function of the way sound is transmitted from the outer to the inner ear, which is known to vary as a non-linear function of frequency. For frequencies below 1 kHz no correction is needed, but above this frequency corrections are needed which will have the effect of increasing the sound level predicted to reach the inner ear. The third stage of the calculation is that of the excitation levels, which involves modelling the functioning of the auditory filter. The fourth stage is the calculation of specific loudness levels, and the fifth involves a calculation of the total loudness considering the excitation levels in the 20 critical bands between 63 Hz and 12.5 kHz. This is done according to an intensity law, with the total amount of loudness being expressed in sones or phones. The excitation pattern of the noise and the excitation pattern of the warning are then superimposed, once all the relevant calculations have been carried out. A warning can be considered to be audible if the excitation pattern of the warning exceeds that of the noise in at least some parts of the spectrum. If the excitation pattern of the noise exceeds that of the warning at all points, the warning will be inaudible. The Detectsound method of predicting the overall level of noise is thus different from that of Patterson's in a number of ways, from the auditory filter model used, to the way in which noise levels and the level at which auditory warnings are to be set are measured.

A discussion of the relative merits of Patterson and Moore's model (1986) and that of Zwicker and Scharf's (1965) which was developed for the Laroche computer model is beyond the scope of this book, but further details can be obtained from the original works (Patterson, 1982; Laroche *et al.*, 1991). The important point is that both models represent vast potential improvements in predicting appropriate levels for auditory warnings in noisy workplaces, partly because they take account of advances in hearing theory – particularly in the fields of frequency selectivity and the auditory filter – and partly because both groups have been instrumental in developing software which allows the calculation of masked threshold and appropriate loudness levels from noise level readings. The logic of both models is much the same, but the output of each is derived in a somewhat different way.

The models proposed by Patterson (1982) and Laroche *et al.* (1991) can tell us whether auditory warnings currently in use are too quiet or too loud, but they are only applicable to relatively steady-state noise. One way to cope with fluctuating noise levels may be to introduce time constants into the calculations, although how this might be implemented is a complex question. Laroche suggests that the solution might be to consider only the maximum noise level where an auditory warning might be heard. The problem is then that the recommended sound level for a warning may be excessively loud, although this is obviously preferable in situations where warnings must be heard as a matter of life or death. One relevant workplace might be the oper-

Table 5.1 Operating room alarms in a hospital found not to reach masked threshold in the presence of other masking sounds. The masker is on the left of the table, the alarms masked are shown on the right (from Momtahan et al., 1993)

Masker	Alarm Masked
Anaesthesia ventilator oxygen supply failure	Oxygen analyser, blood coagulation timing system, intravenous infusion pump
Pulse oximeter, ventilator humidifier, cardiac monitor warning, beeper	Intravenous infusion pump
Blood pressure machine	Pulse oximeter, anaesthesia machine caution
Cardiac monitor rate/pressure	Pulse oximeter, intravenous infusion pump
Anaesthesia ventilator disconnect	Ventilator humidifier
Anaesthesia ventilator oxygen supply failure	Oxygen analyser
Intravenous infusion pump	Anaesthesia ventilator low airway pressure and ventilator failure
Orthopaedic drill	Pulse oximeter, oxygen analyser, anaesthesia ventilator high/low pressure, two anaesthesia ventilator humidifiers, anaesthesia machine warning, anaesthesia machine caution, cardiac monitor critical, beeper, two intravenous infusion pumps, anaesthesia ventilator low airway pressure, anaesthesia ventilator low oxygen, anaesthesia ventilator set volume not delivered
Cast cutter	Same alarms masked by the above drill
Dental drill	Oxygen analyser, intravenous infusion pump
OR5 ambient noise	Anaesthesia ventilator humidifier

ating theatre of a hospital. In such environments it may be necessary to speculate on a 'worst case' scenario; that is, where noisy equipment is being used and where several alarms sound at once. Such speculation should of course lead to consideration of all the other problems associated with the use of auditory warnings, as reviewed in Chapter 4.

A study by Momtahan et al. (1993) gives us some insight into this 'worst case' situation, and also shows how the Detectsound model can be used in determining the detectability of warnings. This study also highlights the kinds of problems which may accompany any prediction of masking which is likely to occur in a fluctuating noise environment. In this study, the audibility and detectability of 23 auditory warnings used in the intensive care unit and the 26 warnings used in the operating rooms of a 214-bed Canadian teaching hospital were assessed. The recognition of the warnings showed some interesting results. From digital tape recordings of the warnings (all 49 of them), the

Detectsound software was used to determine if each of the warnings would be detectable in the noise present at the time of the study, and also if they would be masked by other warnings. Other noisy pieces of machinery and equipment were also taken into account.

Some warnings were excluded from the analysis – for example, intermittent sounds would not be expected to mask continuous sounds and alarms that are never in the same room together were excluded. Tables 5.1 and 5.2 show the operating theatre alarms that did not reach even masked threshold – let alone appropriate detectability levels which would necessarily be about 15 dB higher – when compared with those warnings and other sounds considered to be potential maskers. The item producing the greatest amount of masking was the orthopaedic drill. Some warnings completely masked other warnings, and the authors give the example of a pulse oximeter alarm which, if sounding whilst a cardiac monitor is also on would be completely inaudible because the sound level produced by the pulse oximeter is lower than that of the cardiac monitor at every point in the spectrum (Figure 5.4).

Momtahan et al.'s (1993) study does not address the probability that the alarms and other sounds will actually be heard together, although the authors make some logical and sensible decisions as to whether to include warnings and other sounds on the basis of their structure and their use. In the future this might be established from an observational study or, perhaps less satisfactorily, from retrospective questionnaire studies such as those carried out by O' Carroll (1986) which was discussed in Chapter 4. There are actually two issues involved here. The first is the determination of the likelihood that alarms will sound together, which is a complex and to some extent a fairly intractable problem. Secondly, assuming that there is some probability of masking, there is the question of how the alarm should be designed so that the probability of

Table 5.2 Intensive care unit alarms in a hospital found not to reach threshold in the presence of the listed masking sounds (from Momtahan et al., 1993)

Masker	Alarm masked
Ventilator humidifier	Ventilator high/low pressure
Wall blender	Two ventilator blenders, ventilator high/low pressure, ventilator disconnect, ventilator power failure
Blood pressure machine	Two pulse oximeters, intracranial pressure monitor high/low pressure and airleak
Blood pressure machine	Intracranial pressure monitor airleak
Warming blanket	Ventilator blender, intracranial pressure monitor airleak
Radiant warmer	Ventilator disconnect, ventilator failure
Ambient noise	Ventilator blender, intracranial pressure monitor high/low pressure and airleak
Computer printer	Intravenous infusion pump

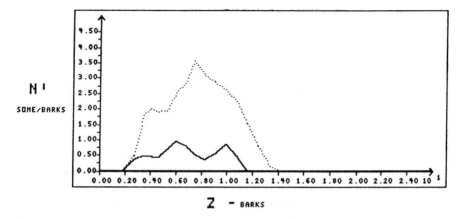

Figure 5.4 Predicted excitation patterns of a cardiac rate/pressure alarm (dotted line) and a pulse oximeter alarm (solid line) based on 'Detectsound' modelling (from Momtahan *et al.*, 1993).

masking is reduced. The necessary steps to achieve this can in some cases be quite simple and will be considered in the next main section on auditory warning design.

In Momtahan *et al.*'s (1993) study the authors found that not only were many of the alarm sounds potentially maskable by others, they were often unpleasant, confusable and hard to localise. For the moment, let us consider the implications of the findings regarding potential masking. First of all, if an orthopaedic drill is being used, then not many of the other warnings will be audible. In addition, some alarms will mask others. Without a complex and lengthy task analysis, or the compiling of a typical 'audio picture' of an operating room, it is very difficult to predict which warnings are likely to be heard together. Even if such masking is predictable it is difficult to see what kinds of solutions might be available.

There are three partial solutions, however, which might assist. The first is to reduce the number of warnings, to reduce the likelihood that alarms will sound together. The second is to centralise warnings so that prioritisation occurs. There are two possible directions here. One, which is beginning to be applied in many spheres where warnings are used, involves the use of some kind of intelligent system which carries out provisional analysis of the input signals that would normally trigger each alarm. The second involves the use of pre-programmed warnings designed to alert when specific situations arise, but which are prioritised from the outset. A third option would be to improve the detectability of each of the warnings by more ergonomic design. For example, a warning sound with one or two very loud harmonics (like the BAC 1–11 warning shown in Figure 5.2) is much less detectable, at the same level, than a warning with several acoustically related harmonics and a complex acoustic structure. This has been demonstrated for the hospital warnings designed by

Stanford *et al.* (1985, 1988), who have shown that a signal-to-noise ratio of − 24 dB still allows certain sorts of sounds to be reliably detected. Furthermore, the masking of these warnings by other pieces of equipment cannot be ruled out. If people are moving about then differential masking will occur depending upon where the person is currently standing.

The problems with variably noisy environments thus seem somewhat insoluble. To average the noise levels and use this as the guideline level is to sidestep the problem, but to fix on the 'worst case' situation, which may be acceptable for a fixed noise environment, would be likely to lead to excessive noise levels and thus excessive warning sound levels. Since it is the simultaneous or near-simultaneous onset and duration of warnings and other sounds which is the source of the most intractable problems it would seem that prioritisation and number reduction are the most obvious solutions. By carrying out both of these the potential for two warnings to mask one another would be reduced and noise levels could be kept down. Psychologically, confusions would also be less likely to occur, so there are yet further advantages to be had from such a move. A relevant research programme for the future would be to combine work on the masking of warnings with that of the study of work patterns and alarm use. Such a research combination could lead to valuable insights concerning the design and use of warnings.

5.2 Warning design

5.2.1 Acoustic and psychoacoustic aspects

Many aspects of warning design relate to cognitive issues such as the reduction of confusion, enhancing the learning of warnings, and the concept of urgency mapping, which are dealt with in the following section. However, a fundamental part of the design of an ergonomic auditory warning is the design of its spectral quality and content, as this will affect a number of more low-level, perceptual attributes of the warning which the hearer would be oblivious to at a conscious level.

Two approaches to auditory warning design which address the spectral content of warnings are discussed here. They are the approaches taken by Patterson (1982), and Stanford *et al.* (1985, 1988). Both of these methods of auditory warning construction propose that the basic spectrum of the warning should be acoustically complex (i.e. the warning should have many, rather than just one or two, harmonic components). This creates two important advantages over warnings relying on just one or two harmonics – they will be easier to localise and be more resistant to masking. Thus the chances are increased that the warning will be both easier to hear and the location where it is coming from will be easier to pinpoint.

Although there are particular problems associated with listening to sounds in rooms where many objects are present, because the sound waves tend to get reflected and deflected, the ears use basically two mechanisms for locating the source of a sound. Both mechanisms capitalise on the fact that we have two ears, with a solid object in between.

At low frequencies, sound waves usually hit each of the ears at a different part of their cycle because there are relatively few cycles per second and the sound waves are large. This phase difference usually allows us to locate the source of a sound. At high frequencies, this information is of little use because the waves cycle much faster, and are consequently much smaller. In high frequency sounds an 'acoustic shadow' is produced by the head and is cast onto the ear furthest from the source. This acoustic shadow produces a drop in level which then becomes the information used to locate the source of the sound. There is a frequency region, however, for which neither mechanism functions very well, and this tends to be around the region of greatest sensitivity, at about 1 kHz. By a strange quirk of fate this is the region in which many auditory warnings are placed, especially the aversive continuous tones often used by hospital equipment. The result is that these tones are difficult to localise, as well as being confusing and aversive.

Improving the localisation quality of an auditory warning can also be helped by making the warning acoustically complex. That is, by giving it a number of harmonics rather than relying on only one or two overly loud harmonics. Because of their irritating qualities, using very high frequencies would not be a good idea so it is most ergonomic to use a fairly low fundamental frequency with a set of associated harmonics. This precise course of action is recommended by Patterson (1982), and can be found in a range of other auditory warning design documents (Lower et al., 1986; Simpson, 1991).

The second advantage in creating an auditory warning with a rich harmonic structure is that the risk of masking will be reduced. If one or two harmonics are momentarily masked, those remaining may still be audible. Even if the fundamental frequency is masked, the pitch of the sound will still be discernible due to the ability of our higher-level cognitive structures to 'fill in' missing information. The ability of the auditory system to hear a fundamental from a set of harmonics, even when it is not present, is a well-known psychoacoustic phenomenon known as the phenomenon of the 'missing fundamental' (e.g. Schouten, 1940; Licklider, 1956).

Let us turn now to the specific design proposals made by Patterson and by Stanford et al. (1985). They differ quite distinctively overall, but the rich harmonic structure of the basic building blocks of the warning sound is common to both.

Patterson's method of construction is shown in Figure 5.5. The basic building block of the warning is a short pulse of sound, lasting usually from 150–300 ms in length. This pulse has an onset and an offset envelope (to avoid startle), and a number of harmonics, at least four of which should be below 4 kHz. Generally, the fundamental should be considerably lower than 1 kHz to improve localisation and reduce aversiveness. Usually the harmonics would be integer multiples of the fundamental frequency (to aid resistance to masking,

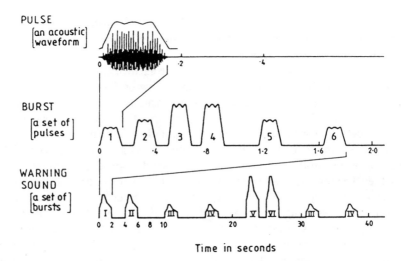

Figure 5.5 Patterson's method of alarm construction. The pulse (top line) is the basic unit of the burst (middle line), which in turn is the basic unit of the warning (bottom line). Time is shown along the bottom in each case (from Lower et al., 1986).

and the perception of a definite pitch) but this is not essential. Thus the pulse of sound contains all the acoustic information. The pulse is then played several times to form a burst of sound. The burst consists of a set of perhaps five or six pulses, lasting from one to three seconds. The pulse is kept constant throughout, but its pitch, and the time intervals between each successive playing of the pulse, can vary. Thus a burst can sound something like an atonal melody with a distinctive rhythm, played on a single instrument. A complete warning is then assembled by playing the burst several times. Perceived urgency can be manipulated by altering the pitch, level, and speed of the burst. The overall warning itself would vary in its precise structure depending upon the priority of the task which it is signalling. An example of the use of warnings like these was discussed at length in Chapter 4.

Patterson's (1982) proposed method of construction solves many of the problems associated with auditory warnings. As well as setting warnings at an appropriate level, which is the first part of his guidance, the pulse itself provides greater resistance to masking and improves the chances of accurate localisation. The construction of bursts of sound and assembling these into complete warnings helps one begin to address many of the more cognitive issues such as confusability and perceived urgency. These elements of warning design have been researched at some length since Patterson produced his guidelines, and will be considered in the following pages.

A second method of warning construction, resulting in warning sounds that are even more novel than those proposed by Patterson, has been proposed by Stanford et al. (1985). Again, the spectrum of the sounds are harmonically rich and in this case the basic building blocks are vowel-like segments formed into a frequency modulated series. Examples of the segments are shown in the top part of Table 5.3 and the temporal characteristics of examples of warnings of this type are shown in the bottom part of the table. In many ways this method of construction is similar to that of Patterson, consisting as it does of a small unit built up into larger chunks.

Both methods of warning construction are radical alternatives to traditional methods of warning construction, but in both cases research has shown that in addition to providing more useful acoustic information (which will govern our subconscious responses to these warnings) there are psychological and aesthetic advantages, which are considered in the next section.

Table 5.3 Acoustic characteristics of Stanford and McIntyre's experimental alarm sounds. (a) shows the spectral characteristics of the vowel-like components (b) shows the arrangements of vowel-like segments and silent internals (s) in the sounds (from Stanford and McIntyre, 1988)

(a)

Component	Fundamental	Formants			
low/i/	(F0 = 100 Hz)	F1 = 300	F2 = 2400	F3 = 3300	F4 = 3900 Hz
high/i/	(F0 = 150 Hz)	F1 = 300	F2 = 2550	F3 = 3450	F4 = 4050 Hz
low/a/	(F0 = 100 Hz)	F1 = 600	F2 = 1200	F3 = 2400	F4 = 3300 Hz
high/a/	(F0 = 150 Hz)	F1 = 600	F2 = 1350	F3 = 2550	F4 = 3300 Hz
low/u/	(F0 = 100 Hz)	F1 = 300	F2 = 600	F3 = 2400	F4 = 3300 Hz
high/u/	(F0 = 150 Hz)	F1 = 300	F2 = 750	F3 = 2550	F4 = 3450 Hz

(b)

Signal no.	Segments
1	$\frac{low/a/}{50\,msec} + \frac{high/i/}{100} + \frac{low/a/}{100} + \frac{high/i/}{100} + \frac{low/a/}{100} + \frac{high/i/}{50}$
2	$\frac{low/a/}{50} + \frac{high/u/}{100} + \frac{low/a/}{100} + \frac{high/u/}{100} + \frac{low/a/}{100} + \frac{high/u/}{50}$
3	$\frac{low/i/}{50} + \frac{high/u/}{100} + \frac{low/i/}{100} + \frac{high/u/}{100} + \frac{low/i/}{100} + \frac{high/u/}{50}$
4	$\frac{high/i/}{50} + \frac{high/u/}{100} + \frac{S}{50} + \frac{low/a/}{100} + \frac{S}{50} + \frac{high/u/}{100} + \frac{S}{50}$
5	$\frac{high/a/}{100} + \frac{high/a/}{100} + \frac{S}{50} + \frac{low/u/}{100} + \frac{high/i/}{100} + \frac{S}{50}$
6	$\frac{high/a/}{100} + \frac{S}{50} + \frac{low/i/}{100} + \frac{S}{50} + \frac{high/u/}{100} + \frac{S}{50} + \frac{low/a/}{50}$
7	$\frac{high/a/}{50} + \frac{S}{50} + \frac{high/i/}{50} + \frac{S}{50} + \frac{high/u/}{50} + \frac{S}{50} + \frac{low/a/}{150} + \frac{S}{50}$
8	$\frac{low/i/}{100} + \frac{low/a/}{100} + \frac{S}{50} + \frac{high/a/}{100} + \frac{low/a/}{50} + \frac{high/i/}{50} + \frac{S}{50}$

5.2.2   Cognitive aspects of warning design

If we are given a verbal warning, we can know what it means, and we stand a good chance of not confusing it with other warnings. We may not comply with it for other reasons, and indeed it may not necessarily be in our best interests to comply with it, as we discussed in Chapter 1. Neither ease of understanding nor lack of confusability are necessarily true of nonverbal auditory warnings because they are essentially iconic in nature, although they can become informational through learning.

If we were to take a hypothetical environment for which no auditory warnings were currently used and were to generate a set of auditory warnings for that environment, one of our primary remits would be to ensure that the total number of warnings did not exceed about one dozen, and users of the system were properly trained in the recognition of the warning sounds. We could then think about ways in which the system could be rationalised so that the number of warnings would not begin to mushroom. One useful method would be to prioritise, and use individual warnings for each of the top priority situations only. A general annunciator could be used for other priorities, because there would be time for the receiver to check on other, more detailed information. This was precisely the regime used in the generation of helicopter warning sounds, as discussed in Chapter 4.

Unfortunately the discipline of Human Factors does not operate in a vacuum, but is subject to the same political and financial pressures as any other line of activity. Thus reductions in the numbers of alarms used in a system is almost bound to be met with resistance because of fears for reduced safety, or the possibility of litigation, even if ergonomically this is for the best. One can also imagine that the purseholder funding a complete rationalisation of a warning system is not going to consider that he or she is getting much of a bargain if, after having spent a great deal of money, there remain fewer warnings than before. Thus the most simple problem associated with auditory warnings – their excessive number – is, paradoxically, the hardest to solve, unless some kind of corporate will is involved. Human Factors experts can, however, have an input to the two other major cognitive aspects of auditory warning design. These are the improvement in the portrayal of appropriate meaning, and the reduction of confusion.

5.2.2.1   Meaning

The meaning of a non-verbal auditory warning is a relatively more important question within the whole process of design than it is for verbal warnings, which are associated with other problems even when their meaning may be clear. The balance between the iconic and the informational is different for these two types of warning. It would be useful, therefore, if we could capitalise on what people already know about auditory warnings and their meanings. For example, it may be true that well-known auditory warnings of the traditional type mean approximately the same thing to the majority of listeners. An example of this type of sound would be a fire bell, which most people recognise. We should, however, guard against using known sounds for new situations. If sounds with learned associations must be used, they should be used for the situations where they are typically known rather than being assigned new meanings.

Psychologically the meanings typically associated with traditional warning sounds are generally learnt rather than being innate, although there may be some psychoacoustic features of sounds like sirens which provoke somewhat basic responses. Some very useful research on these learned associations has been carried out in Germany and Japan (Lazarus and Hoge, 1986; Hoge *et al.*, 1988). Some of this research was carried out on traditional bells, horns, sirens and buzzers, but some more recent impulsed alarms have been studied, such as the 'Quarte tone', a two-tone sequence used on German police cars.

Lazarus and Hoge (1986) asked 20 subjects to scale a total of 41 alarm sounds, 20 of which were appropriate to warning situations, to 4 sets of 12 situations in which the sounds might be used. The 3 groups of 12 situations which are relevant to warning design were those pertaining to specific danger situations, general danger and pleasant situations (which are inappropriate, although relevant, to warning design). They found that some warnings are better matched to certain situations than are others. This implies that there is some inherent advantage in using particular warning sounds for particular situations in which a warning is needed. The psychological basis for this appropriateness is likely to have been learnt, rather than being an affect of the sound itself, but in terms of warning implementation the ultimate psychological explanation is somewhat irrelevant.

This study showed that sirens were rated as being the best fit to Danger or Threat categories, and especially with those associated with categories covering stress and 'Danger to Life'. Bells were found to be the least good fit to the categories generally. Horns fitted most appropriately into categories associated with specific warnings related to machinery.

A cross-cultural study (Hoge *et al.*, 1988) required subjects in both Germany and Japan to rate 20 danger signals on 12 semantic differential scales, and 5 additional scales. The actual warnings used were similar to those used by Lazarus and Hoge (1986) and again consisted of various types of bells, sirens, horns and impulsed sounds. The warnings were scaled on 17 scales and were broken down into seven categories – 'Tendency for Action', 'Perceived Loudness', 'Perceived Pitch', 'Dangerousness', 'Activity', 'Evaluation' and 'Potency'. The results showed that some of the warnings produced significant differences between the two ethnic groupings for the seven categories, but that some of the responses were remarkably consistent. The most obvious of these was on the 'Tendency for Action' scale, which produced the smallest number of differences between the cultural groups. Amongst the sounds, the sirens produced the most consistent meaning between the two cultural groups. Sirens also tended to produce higher means from the rating scales, suggesting that they are generally more suitable as warning sounds, as the earlier culturally specific study showed.

This research thus demonstrates a general appropriateness of sirens for danger situations, an appropriateness of horns for mechanically-related warnings, and greater disagreement about the use of bells. These associations could be exploited in the design of future auditory warning sets. The warnings which were studied were of the traditional type, possessing many of the undesirable qualities associated with such sounds, such as being unnecessarily continuous, aversive, and speech- and thought-interrupting. However, it is possible to synthesise siren-like, horn-like, or bell-like warning sounds which might be much more ergonomically acceptable. In doing this, the warnings would maintain the learned associations but lose their adverse characteristics. Another interesting point about the cross-cultural study was that the response to sirens was more homogeneous between the two groups than was the response to other types of sound. Two possible explanations can be given for this finding. The first explanation is that sirens are used more universally than the other alarms, and so German and Japanese people have the same learned reaction and response to them. The second explanation, the more intriguing of the two, is that there may be something within the sound of the siren itself which provokes the response obtained, and which to some extent is innate. This idea gains some support when we recall that the responses were, in general, most consistent across the two groups in the 'Tendency for Action' category. This might suggest that certain sound parameters have the ability to convey certain things to us without having to learn them, and can suggest that we should do something, albeit nonspecific. For example, the rising pitch of a typical siren may perhaps convey a certain affect of movement without having to learn this.

The idea that sounds, especially music, have 'embodied meaning' (Meyer, 1956) could provide a rich source of ideas for auditory warning design because it has the attraction that the subconscious or unconscious associations between certain types of sounds and certain types of affect or emotion do not need to be learnt. A similar idea is encapsulated in Gibson's idea of 'affordances' (1979). Furthermore, the sorts of sounds which are good at invoking such images or responses might be better retained even when the receiver is under a high degree of stress. Again, the psychological origin of this phenomenon is rather irrelevant for the purposes of design, but is interesting in itself. Nowhere are the almost innate evocations of sounds better exploited than in film music, where the mood of a scene can be greatly enhanced by an appropriate sound track or completely ruined by one which is inappropriate. We may also like to consider for a moment that hardly any culture uses fast, jolly music at a funeral. In the rare instances where such music is used it is because the connotations of the recent death and the way that a particular culture deals with it is different from, and more positive than, our own.

5.2.2.2 Perceived Urgency

In our experience, two of the criticisms often levelled at an auditory warning is that it is either too urgent or not urgent enough. One interpretation of this complaint is that, on the evidence of the situation that the warning is meant to be representing, the opinion is that the auditory warning icon for that situation is inappropriate. Thus users are implicitly advocating some system of urgency mapping. For example, many of the warnings used in hospitals are shrill and alarming when they do not need to be, and medical staff often comment informally that the warnings are unnecessarily urgent for many applications. The study by O'Carroll (1986) cited in Chapter 4 showed that of 1455 soundings of alarms over a 3-week period only eight represented life-threatening situations. All of the other situations were nonlife-threatening, but were urgent in varying degrees. Studies of alarm systems in hospitals have also shown also that medical staff working in a particular environment show a lack of knowledge of the meaning of many of the alarms that they might typically hear. These two points suggest that since there are many alarms and the receivers tend not to know their meanings, it may be useful if each alarm conveyed at least something of the appropriate urgency required of the response to it. The obvious way to do this would be to attempt to convey at least the urgency of the situation through the warning sound itself.

It is necessary to make two assumptions here. The first is that urgency is an attribute of sound, and an evaluation of the urgency of a sound can be made even when the listener does not know the meaning of the warning, because urgency is a function of the mix of acoustic parameters and the intensity of these parameters in a warning. The second assumption is that up to a point the urgency of certain situations is context-independent. Of course there are situations where urgency is indeed context-dependent, but with appropriate use of artificial intelligence there is no reason why, in time, context-dependent urgency could not also be appropriately conveyed through changeable auditory warnings.

The concept of perceived urgency is important both in the hospital and helicopter developments highlighted in Chapter 4. In the helicopter environment the priorities are generally context-independent – the priority of different situations is fixed by the red, amber and green light system and the urgency of the auditory warnings which go with these three categories is correspondingly graded. As Chapter 4 shows, urgency mapping is becoming an important part of hospital alarm development and is embodied within British, European, US and worldwide standards.

Given that urgency mapping might be a useful improvement to an auditory warning system, it is essential to know how to manipulate sound parameters in order to produce urgency contrasts. This has been done in a series of studies (Momtahan, 1990; Edworthy et al., 1991; Hellier et al., 1993; summarised in Edworthy, 1994). Many of the findings replicate our preconceptions about how the urgency of sound is affected by changes in that sound, but some findings are rather more counter-intuitive.

Patterson's (1982) method of auditory warning construction (Figure 5.5) lends itself well to the concept of urgency mapping. Many traditional warnings are continuous mechanically-produced sounds, meaning that they cannot be altered from their original form except perhaps to increase their loudness or pitch, both of which may make the warnings unacceptably aversive. If, on the other hand, warnings are produced by computer and are programmed into chips, not only can the warnings sound exactly as we would want them to sound, but many of their features can be altered at will in the search for the appropriately urgent version of that warning. Patterson's method of warning construction starts off with a single pulse of sound containing all the necessary acoustic information. A burst of sound is created by playing the pulse several times, possibly at different pitches and with different time intervals between them. As these bursts are somewhat akin to atonal melodies with a distinctive rhythm, many ways of altering the urgency of these warnings is possible. In one set of studies (Edworthy *et al.*, 1991) the effects first of pulse parameters, and then of burst parameters, on the perceived urgency of rather simple auditory stimuli were systematically charted. The pulse parameters included temporal features such as the onset and offset amplitude envelope of the pulse (the rate at which the full power of the pulse is reached after its onset), harmonic features such as the fundamental frequency of the pulse, and the relationship of the harmonics to one another. In most musical tones the relationship between the fundamental frequency, which gives a note its pitch, and the harmonics above it, which give it its timbre, is that of a simple integer. That is, if the fundamental is 200 Hz the harmonics will all be multiples of 200 although not necessarily all of them need be present. Many naturally occurring sounds, and certainly those that can be produced by computer, do not retain this simple integer relationship and thus connote a degree of inharmonicity. The studies referred to also considered relatively simple forms of such inharmonicity, where all of the harmonic components of a specific sound were either 10 or 50 per cent above or below their integer multiple value, or were completely random. The experimental paradigm adopted was one of multiple comparison (Gulliksen and Tucker, 1961) which meant that all stimuli were compared with every other, twice. The results obtained are shown in Table 5.4.

The results showed only the direction of these effects, not their precise strengths. For example, although the results confirmed our intuition by

Table 5.4 Effects of four pulse characteristics on perceived urgency (from Edworthy, 1994)

Parameter	Direction of effect
Fundamental frequency	High > low
Harmonic series	Random/10% irregular > 50% irregular > regular
Delayed harmonics	No delayed harmonics > delayed harmonics
Amplitude envelope	Regular/slow onset > slow offset

Key: > More urgent than
      / Equally urgent

showing quite unequivocally that higher pitched sounds are more urgent than lower pitched sounds and that inharmonic sounds are more urgent than harmonic ones, the results alone do not show which of those two parameters has the stronger effect on perceived urgency. That question was tackled in later studies.

In the second part of the first study the effects of burst parameters on perceived urgency were charted. These are the acoustic parameters which affect the temporal and melodic quality of the longer unit of sound, made up from several individual pulses. The parameters studied and the main effects found can be seen in Table 5.5. These results showed that speed (as measured by pulse-to-pulse interval) affects urgency, which confirmed Patterson's earlier recommendation (1982). Other temporal features such as rhythm and simply the number of times a burst is repeated also affect its urgency. Amongst the melodic parameters it was found that the pitch range (the difference in pitch from the highest to the lowest pitch in the burst) affects urgency, as does pitch contour (the pattern of ups and downs in a burst) to a lesser extent. Even the melodic structure of the burst affects its urgency. A musical, resolved stimulus was rated as less urgent than an unresolved one, which in turn was rated as less urgent than one which was atonal. These effects varied in their strength, although precise details could not be derived from this study alone.

There was a high level of agreement about the direction of these effects, and it was found that subjects had no difficulty at all in making urgency judgements about these relatively simple acoustic stimuli. There was very strong inter-subject agreement, and for every single parameter tested a significant effect was obtained across the range of levels used. This suggests that perceived urgency is an accessible feature of many kinds of warning parameter, and that the strengths of the effects are substantial. In the final experiment in this series a set of 13 warnings was designed which were predicted to vary in their degree of urgency. These are shown in Table 5.6. These warnings were generated by creating pulses with expected high urgency by combining highly rated levels of different parameters (such as, for example, a high fundamental frequency, a set of irregular harmonics and a regular or slow onset amplitude envelope) and by turning these into what were expected to be highly urgent bursts by using

Table 5.5 Effects of seven burst characteristics on perceived urgency (from Edworthy, 1994)

Parameter	Direction of effect
Speed	Fast > moderate > slow
Number of repeating units	4 > 2 > 1
Rhythm	Regular > syncopated
Speed change	Speeding up > regular/slowing
Pitch contour	Random > down/up
Pitch range	Large > small > moderate
Musical structure	Atonal > unresolved > resolved

Key: > More urgent than

Table 5.6　Compositions tested by Edworthy et al. (1991)

Warning	Pulse characteristics			Burst characteristics			
	Envelope	Harmonic regularity	Pulse-Pulse interval	Rhythm	Average pitch	Pitch range	Pitch contour
1	Standard	Random	150 ms	Regular	600 Hz	300 Hz	Random
2	Standard	10% Irregular	175 ms	Regular	585 Hz	230 Hz	Random
3	Slow onset	Random	200 ms	Speeding	510 Hz	280 Hz	Random
4	Slow onset	10% Irregular	225 ms	Regular	525 Hz	200 Hz	Up/down/up
5	Standard	50% Irregular	250 ms	Syncopated	500 Hz	275 Hz	Down/up × 2
6	Standard	10% Irregular	275 ms	Speeding	450 Hz	100 Hz	Up/down/up
7	Standard	50% Irregular	300 ms	Syncopated	400 Hz	125 Hz	Up/down
8	Slow offset	50% Irregular	325 ms	Regular	335 Hz	170 Hz	Down/up
9	Slow offset	Regular	350 ms	Syncopated	300 Hz	120 Hz	Up
10	Slow offset	Regular	375 ms	Regular	250 Hz	75 Hz	Down/up
11	Slow offset	Regular	400 ms	Syncopated	210 Hz	80 Hz	Down
12	Slow onset	Regular	450 ms	Slowing	175 Hz	50 Hz	Up/down
13	Slow onset	Regular	550 ms	Slowing	290 Hz	75 Hz	Down

these pulses as the basis for bursts with highly urgent levels of a range of parameters (such as, for example, a fast speed, a large pitch range, an atonal pitch pattern, and perhaps by repeating the burst). The converse was done for expected low urgency pulses and bursts, and a range was created by systematically varying the urgency of the pulses and bursts between the two extremes (for details, see Edworthy *et al.*, 1991). The same multiple comparisons procedure as before was then carried out on this set of warnings, and the correlation between the predicted and the obtained order proved to be highly significant. These results show, therefore, that perceived urgency is predictable and can be manipulated.

Although this study of perceived urgency focused on sound parameters which are particularly relevant to the design of warning sounds similar to those proposed by Patterson (1982), there is no reason why the results should not generalise to warnings of other types, provided that the parameter is meaningful with relation to that warning. For example, the pulse parameters which affect the spectrum could equally affect the spectrum of any warning, traditional or otherwise. Although the earliest work suggested that the factors which affect the urgency of the pulse are less important to overall urgency than those which affect the urgency of the burst (Edworthy *et al.*, 1988), it is likely that spectral parameters exert greater influence on more traditional types of warnings, in which features such as pitch, warning speed and so on are not as relevant. If, however, traditional types of warnings (such as a siren, for example) are resynthesised according to Patterson's principles, then its urgency can be manipulated to a far greater extent. The advantage of such a practice is that the sound already has some learned associations – thus in a sense we can have the best of both worlds here.

The effects of other important warning parameters on perceived urgency have been investigated by Momtahan (1990). Momtahan's methods, although similar to those used by Edworthy *et al.* (1991), required subjects to make comparisons between levels of sound parameters using a paired rather than a multiple comparison task. Momtahan investigated pulse length, interpulse interval, amplitude modulation, number of harmonics, spectral shape, fundamental frequency, frequency modulation, number and direction of frequency glides, amplitude envelope shape and number and direction of pitch steps. These parameters are described in Table 5.7. They are therefore a mixture of pulse and burst parameters among which are some parameters not included in the Edworthy study but which are nevertheless likely to be important in the perception of urgency. Some of the parameters would figure quite strongly in warnings designed in the style of electronic computer game sounds rather than in the style of the more temporally oriented warnings advocated by Patterson (1982). It is difficult, because of the relatively short length of the pulse. to include features such as pitch glides, amplitude modulation and frequency modulation in bursts of the Patterson variety. The second experiment in Momtahan's (1990) study also included intensity, an obvious component of perceived urgency but one ignored in the Edworthy studies because, in many

Table 5.7 Stimuli tested in Momtahen's studies (1990)

Parameter	Description	Results
Pulse length	100, 250, 500 ms	100 > 250 > 500 ms
Interpulse interval	100, 250, 500 ms	100 > 250 > 500 ms
Amplitude modulation	no modulation, 4, 8, 16 Hz	no mod > 4, 8, 16 Hz
Number of harmonics	fo only, fo + 3 harmonics, fo + 7, fo + 11 harmonics	11 > 7 > 3 > fo only
Spectral shape	equal energy across all harmonics, stepwise, majority of energy in middle, maj energy in lower and higher harmonics, bias towards last 3 harmonics, square wave	equal/low-high > rest
Fundamental frequency	143, 330, 500, 660, 1000 Hz	1000 > 660 > 500 > 330 > 143 Hz
Frequency modulation	no fm, 4, 8, 16 Hz	none > 4, 8, 16 Hz
Frequency glides	none, 350–400, 400–350, 350–400–350, 400, 350, 400, 350–400–350–400, 400, 350, 400, 350 Hz	none > all glides (except up)
Amplitude envelope	rise times of 25, 100, 200 ms, decay times of 25, 100, 200 ms	Rise: 25 > 100, 200
		Decay: 25 > 100, 200
Pitch steps	375, 350–375–350, 350–375–400–375–350, 375–350–375, 400–375–350–375–400 Hz	375 > all pitch steps (those ending with steps going up more urgent than those ending with down step)

environments, the loudness of a stimulus will be determined and fixed by the background noise although it could vary within the 15–25 dB above-threshold range.

In Momtahan's study subjects were presented with 3-second windows of sound and all possible pairings of the stimuli were heard. In each case, subjects had to judge which of the two stimuli was more urgent. The results showed that for some stimuli there was no consistent trend in rating one level as more urgent than another, or in a consistent direction. The results largely replicated Edworthy *et al.*'s (1991) findings for those parameters which were similar to those tested by them, such as interpulse interval, fundamental frequency, and pitch contour (pitch steps in Momtahan's terminology). There were also some contrasts, such as the effect found for amplitude envelope. Some of the other parameters also showed interesting effects. For example, the absence of amplitude modulation was found to be more urgent than the presence of amplitude modulation. A greater number of harmonics was found to be more urgent than fewer. Momtahan (1990) also found that sound pulses with harmonics of equal weight, or more energy in the lower and higher harmonics (i.e. not the middle) were judged to be more urgent than a range of other possible spectral shapes. Another finding was that the presence of frequency modulation was judged to be less urgent than the absence of frequency modulation, and that in general the absence of a frequency glide was judged to be more urgent than its presence, although up-glides were judged as being more urgent than down-glides, or up/down glides. Momtahan's results are also summarised in Table 5.7.

In a second experiment Momtahan (1990) took the most successful (success defined in terms of the consistency with which individual stimuli were rated as being more urgent than the one with which they were paired) of these parameters – interpulse interval, number of harmonics, spectral shape, fundamental frequency, frequency glide and loudness – combined them, and assessed the contribution of each of them to the overall judgement of urgency by measuring the amount of variance accounted for in the data obtained using a 'low' and a 'high' urgency version of each parameter. The pairs of stimuli for each parameter were taken from the two ends of the range of each variable under consideration. The results for each individual parameter are shown in Figure 5.6. Momtahan found that all six parameters produced main effects, along with a range of interactions. She found that loudness had the largest effect, followed by interpulse interval. More recently, Haas and Casali (1995) have replicated the finding that intensity has a greater effect on perceived urgency than other acoustic parameters. However, the ergonomically-useful range over which loudness may vary is fairly small, because of the adverse effects associated with having warnings too much louder than threshold. Other factors such as warning speed can be allowed to vary over a greater range without producing adverse side-effects, so these are likely to be more useful in the manipulation of perceived urgency in practice. Interestingly, Momtahan's study did not show a clear effect for fundamental frequency, which otherwise shows fairly

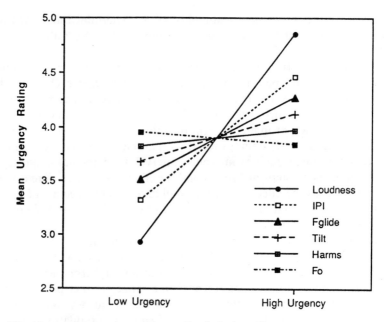

Figure 5.6   Mean urgency ratings of the two levels (high and low urgency) of the six sound parameters tested by Momtahan (from Momtahan, 1990).

clear and consistent effects in other studies. Momtahan's method of directly comparing six parameters in which the range was more or less arbitrarily chosen is to some extent a problem in terms of generalisability and calibration, but the study demonstrates many interesting effects.

A number of studies have also been carried out investigating the measurement and quantification of perceived urgency (Hellier and Edworthy, 1989; Hellier *et al.*, 1993). Momtahan's study points to the observation that some parameters have stronger effects on urgency than others, and this was also implicit from the Edworthy studies. One reason why it might be useful to ascertain the relative strengths of these effects is that if urgency mapping is to be attained in a specific set of warnings, warnings need to be graded in their urgency. It would also be necessary to generate warnings that are different, but which are approximately equally urgent.

We need to know the relative strengths of different auditory warning parameters on urgency to propose ways in which warnings can be constructed in such a way as to be reliably different from one another in urgency level. Additionally, we need to know how to make warnings of approximately equal urgency which sound different from one another, so that confusions do not occur. However, a composite auditory warning consisting of various levels of different parameters such as level, speed and pitch cannot be broken down into its constituent parts in the same way as for a warning label (where for example, we can readily distinguish signal word from colour and from font size). Thus a technique rather different from the isoperformance curves of Braun *et al.* (1994) may need to be used.

One technique which allows us to look at single dimensions of a stimulus in a reliable way, with a view to later combining these measurements, is the use of psychophysical functions. One well-used method of quantifying relationships between the physical intensity of a stimulus (such as its loudness, brightness, or weight) and subjective judgement is Stevens' Power Law (1957), which states that

$$S = KO^m$$

where S is the subjective value given to a stimulus, K is a constant, O is the objective and measurable value of some physical continuum and m is the exponent, revealed when subjective and objective values are plotted against one another and the result is fitted by a power function. Although Stevens' Power Law has been widely used for assessing the direct relationship between physical and subjective intensity (like judging the subjective loudness, brightness or heaviness of a stimulus, and comparing this with the physical intensity, luminance or weight of the stimulus), it has rarely been applied to the scaling of subjective phenomena in which a stimulus is varied and the subject asked to rate some second-order feature of that manipulation. This has, however, been done for the annoyance value of noise (Kuwano and Namba, 1990; Hellman and Zwicker, 1990). Before using this method it was ascertained, through a series of experiments, that perceived urgency was measurable using typical magnitude scaling procedures (Hellier *et al.*, 1995). A number of studies followed where the more important, and quantifiable, acoustic parameters known to have large effects on perceived urgency were quantified (Hellier *et al.*, 1993). The main parameters considered were speed, pitch, number of repetitions and degree of inharmonicity. An exponent was derived for each of these which can be seen in Table 5.8.

The higher the exponent, the stronger is the parameter in producing changes in perceived urgency. Large exponents require only small changes in the value of the physical parameter to produce a unit change in urgency, whereas smaller exponents require larger changes to produce the same unit change in assessment of urgency. One of the great advantages of this method of quantification is that the exponents can be used to predict how much

Table 5.8 Urgency exponents for four auditory warning parameters (from Edworthy, 1994)

Parameter	Exponent
Speed	1.35
Fundamental frequency	0.38
Number of repeating units	0.50
Inharmonicity	0.12

change in the physical parameter is required to produce a specified change in perceived urgency. In fact, if the functions are drawn on a graph, equal changes in urgency (and equal levels of urgency) can be read off for different parameters. Alternatively, the values can be calculated using the exponent values. The lower the exponent, the greater is the change needed in that physical parameter to produce a unit change in urgency. For example, changes in inharmonicity produced a very small exponent, so that parameter is of little use in altering urgency. It has to be borne in mind of course that other manipulations of inharmonicity, not explored in the Edworthy studies (for example, Edworthy, 1994; Edworthy et al., 1991) may produce greater changes in urgency.

The other advantage of such quantification is that it may allow us to equate urgency levels across parameters. The power equations can be used to generate equivalent levels of urgency using different acoustic parameters. For example, a speed of 5 pulses per second and a pitch value of 380 Hz are theoretically equally urgent (Hellier et al., 1993). Information such as this is useful in generating auditory warnings which are different, but are approximately equally urgent. More theoretically, we can use this equating of urgency to ascertain whether some acoustic parameters contribute more to judgements of urgency than others, even when they are theoretically equal. One of the problems in the earlier studies was that it was not known which of the parameters was contributing most to the judgement of urgency. This has been a problem in other psychoacoustic studies looking at other meanings in sound. For example, Freed and Martens (1986) carried out a study on the perceived hardness of metallic sounds, looking at the effects of eight individual acoustic parameters, but they had to concede that their results were limited in that the selection of levels was rather arbitrary. It is unwise to say that one acoustic parameter contributes more to a particular subjective judgement (such as urgency, or hardness) than another if the levels of some parameters presented to subjects covered a wider range than others, a point about the calibration of warning parameters and the selection of levels of independent variables which was emphasised in Chapter 1. With equalised urgency levels across different parameters it was found that pitch was contributing more to judgements of urgency than speed and number of repetitions, the other two parameters tested. One possible reason for this is that pitch is a feature of sound where changes in level produce qualitative, rather than quantitative, changes in the sound heard and may thus be more salient as a result. This is an important aspect of discrimination between warnings, which will be dealt with in the next section.

Our own practical experience of developing auditory warnings (for example Lower et al., 1986; Patterson et al., 1986; Edworthy and Hellier, 1992a) suggests that perceived urgency, and urgency mapping, is an important requirement for advanced auditory warnings systems, not least because the acceptance or the rejection of a proposed set of warnings can depend upon whether the warnings are judged to be sufficiently urgent, or sufficiently nonurgent. Although first reactions to warnings are not necessarily good indicators of whether warnings will be effective, if their acoustic characteristics do not commend themselves to the people who are likely to be using the warnings in the future they stand less chance of being taken up in practice. On a more scientific level, the consistency of the responses that was found during these studies, together with the ease and readiness with which most people assess the relative urgencies of warnings, suggests to us that urgency is an important attribute of sound and its separable acoustic parameters. Recent work on warning labels (Adams and Edworthy, 1995) suggests that the same may be true of our responses to warning labels. Here again, cross-parameter comparisons can be made.

The study of perceived urgency has so far proved fruitful. There are, however, a number of important issues which need to be discussed at this point. First, it must be considered if urgency mapping is of any added benefit to a warning system. We have argued throughout this book that one of the main purposes of a warning is to represent the referent in both an iconic and an informational way, and one of the most important things that it can do is to reflect the urgency or importance of the referent. Thus we argue that urgency mapping is essential, otherwise there is no point in having the warning at all. It seems logical that if some attempt at urgency mapping is made, then at least the receiver is given some indication of how quickly he or she is expected to act. Thus, from this point of view alone, urgency mapping should increase warning effectiveness.

A second issue again concerns the external validity of urgency mapping. Even if near-perfect mapping is achieved, are warnings dealt with any more effectively than sets in which such mapping is not achieved? Notwithstanding the argument that at least some meaning is conveyed if mapping is achieved, appropriate urgency mapping need not necessarily increase the probability that appropriate behaviour will occur. It may, however, improve the acceptability of the working environment, and thus improve other work areas not directly related to alarms. This question is ripe for investigation and recent work by Haas and Casali (1995) and Burt et al. (1995) has begun to address this issue. Early results show that simple behavioural measures such as reaction time decrease as the perceived urgency of the warning sound increases.

The question of context-dependent urgency also needs to be discussed. Urgency mapping is possible for fairly fixed priorities, because warnings can be matched to situations on the basis of their typical level of urgency, but it would be more difficult to achieve such mapping where the urgency of situations is context-dependent. However, if artificial intelligence plays a large part in the analysis of information on the status of physical parameters relating to a hospital patient or a plant control room, as it increasingly does, there is no reason why an auditory warning system could not be equally context-dependent. In other words, the computer assimilating the information could provide warnings generated from algorithms which determine the urgency of the warnings presented. This is achievable, and some simple systems like this are now in use (e.g. Schreiber and Schreiber, 1989). However, a complex system of this kind is still probably some way off but, again, this is an area ripe for research and design impetus.

Before turning to other aspects of meaning in sound, one interesting finding that arose both from Edworthy's and Momtahan's work was the contrast between the urgency and the distinctiveness of sounds. In colloquial terms, any manipulation of a sound which was rather unimaginative tended to produce predictable effects on urgency. Anything which could be done to a sound to make it more distinctive, such as the addition of delayed harmonics, frequency modulation, or amplitude modulation, tended to make the warning less urgent when compared with the same stimulus without this adaptation. One interpretation of these findings is that manipulations which make sounds distinctive by modulating some parameter which would otherwise be fixed have a tendency to decrease the urgency of the warning. By the same token, however, these features confer distinctiveness on warnings. Thus urgency and distinctiveness are not the same thing, and in some sense seem even to contradict one another. This causes something of a problem for the designer who, in wanting to make the warnings for certain high-priority situations stand out, attempts to make the warning sound distinctive by doing something imaginative and creative with it. Research work suggests that anything which is too imaginative may be counter-productive, at least in terms of urgency.

Urgency is not the only meaning that can be conveyed through a warning sound, but it is a crucial one. Additional benefit might be derived if warnings could convey not only particular levels of urgency but something about the nature of the situation causing the alarm to sound, and even perhaps the way in which a situational parameter is changing. One way into this is to consolidate already-learned associations between warning sounds and situations, as demonstrated by Lazarus and Hoge's work (1986) described earlier. Another way is to explore the way sounds naturally convey meaning to us, and to capitalise on this in the design of auditory warnings. Recently a user-centred methodology for the design and selection of auditory warning sounds has been proposed (Edworthy and Stanton, 1995) which may allow us to capitalise upon these associations, and possibly open up the range of warning sounds which might be suitable for use. Such data might also be useful in the design of monitoring sounds, which will be considered later.

5.2.2.3   Confusion

Another central psychological problem associated with the typical usage of auditory warnings is that they are confusing. This is a problem shared with nonverbal visual warnings, because the meaning cannot be spelled out in the same way that it can for verbal warnings – although in both cases they can be backed up by verbal information. Auditory warnings may be confusing because there are simply too many of them to remember. Several of them may sound similar to one another, or they may be associated with similar situations. They may never have been properly learned, and they may be inappropriate in some way such as being unnecessarily urgent and badly matched to their referents. If most working environments are anything to go by, the causes for confusion are probably a combination of all of these. The problem of inappropriate urgency levels has been dealt with at length in the previous section, and the problem of the number of warnings has also been considered both earlier in this chapter and in Chapter 4. Inadequate training is an issue which can be fairly easily solved if the will is there, and confusions between situations are a function of the work environment and not of the warning itself, although this problem can be approached indirectly through the warning design process.

It is a truism to say that within a set of auditory warnings each of the warnings should be as distinctive from one another as possible. A problem is that standards and guidelines often make statements about this requirement without telling the designer how such distinctiveness might be achieved. In addition, as auditory warnings tend to be added one at a time to a working environment on a rather ad hoc basis the cause of good discriminability between them is not promoted. A way of starting to look at the problem of confusion may be to examine the acoustic structure of all the warnings in a given environment to see where confusions might lie. This approach was taken by McIntyre and Stanford (1985) with a set of anaesthesia alarms, as described in Chapter 4. Although examination of the acoustic characteristics of a warning sound will tell us some of the more obvious sources of confusion – for example that a 2800 Hz and a 2900 Hz continuous tone will obviously be confused – it does not necessarily tell us when two warnings which look quite different on paper will be reliably confused with one another (Meredith and Edworthy, 1994). This issue of what causes confusions between sounds is complex and to consider it fully the psychological salience of individual warning parameters must be considered.

There is a fairly large body of research on the nature of confusion between sounds. Some of this is theoretical in nature and some is more directly addressed to the issue of warning confusion. Before this research is considered, it is important to note at the outset that very little of this research has addressed what is actually a central issue. This issue concerns the relationship between sounds and their non-auditory representations, particularly the verbal labels which are given to sounds, and the kinds of spatial images they produce. Sounds are not just confused because they are acoustically (or psychoacoustically) similar in some way. They can be confused because they share similar verbal labels even if acoustically they are quite different from one another. They can be confused because they produce similar spatial images and because they have similar functions even though they may be perceptually distinct. There is also the issue of particular confusions within a set of auditory warnings, which to some extent may be dictated by the similarities between individual warnings within the set and the labels which are subsequently used to differentiate between the sounds. For example, in a set of warnings contain-

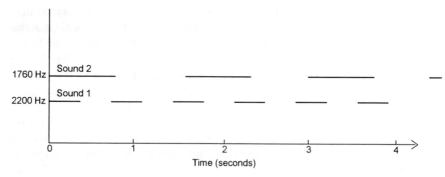

Figure 5.7 Temporal patterns of two auditory warnings found to be readily confused in Meredith and Edworthy's study (1994).

ing only one continuous tone, the label 'continuous tone' will suffice to distinguish this warning from all others. However, if four such warnings are found in that set – such as the one considered by McIntyre and Stanford (1985) – more subtle differentiation will be needed. For example, the user will need to distinguish between high and low tones (if they can), and then find some other, even finer, source of discrimination between warnings which still cannot be differentiated by the pitch distinction. This is an interesting problem and one ripe for investigation at a theoretical level. We will be returning to the more general area of semantic meaning in sound in the section 5.2.3, which concerns monitoring sounds.

Research on confusion between sounds has, perhaps expectedly, tended to focus on acoustic sources of confusion. Auditory perception studies (Webster et al., 1973) show that, at least under controlled experimental conditions, it becomes easier to discriminate between stimuli as the number of acoustic dimensions on which they differ increases. This is not surprising, as we would expect this to happen for any perceptual continuum. However, it is useful to know and apply to auditory warning design. Apart from anything else, this piece of information helps us to open up the design envelope. In the Edworthy (1991) studies of perceived urgency it was found that the final set of 13 warnings designed on the basis of studies which used more simple, less distinctive, stimuli were more discriminable than the simpler, more homogeneous stimuli. Thus, making warnings acoustically and psychoacoustically rich is one way of at least opening up the potential for discriminability between the final warning set, provided that this richness is exploited in creating distinctions between the warnings themselves.

The experimental data most applicable to that of auditory warning design is of course that which has been gathered on auditory warnings themselves. Within this data we find that some parameters of sounds are more salient than others. These more salient parameters will have a large impact on confusion if similar values are shared between warnings, and conversely we would expect differentiation to be aided if these same parameters are varied over a large range. The most fundamental of these seems to be the temporal pattern of the stimulus. Patterson and Milroy (1980) found that warnings which were otherwise quite distinct from one another were readily confused if they shared the same temporal pattern. Warnings could have a totally distinctive spectral structure and yet be easily confused on the basis of their temporal patterns. The first recommendation for making warnings distinctive should therefore be to make them temporally distinctive. More recent studies (Meredith and Edworthy, 1994) have shown that the nature of this temporal confusion extends even to the sharing of on/off patterns which cycle at completely different rates. Figure 5.7 shows the temporal patterns of two warnings which were regularly confused in these studies. It was also found that two warnings which began with a similar pattern (a long tone) and then continued in quite different ways were confused. Table 5.9 shows some of the major sources of confusion found in these studies.

All of this suggests that the temporal patterning of a warning is extremely important, and can be a source of major confusion between warnings. Thus it is wise to keep the temporal patterns of individual warnings within a warning set as distinctive as possible. Some patterns could consist of many short pulses, some could be much longer, but not continuous – one such alarm in a set of alarms would certainly be distinctive, but continuous tones have many adverse characteristics associated with them. Up to a point, all of the warnings could have the same spectrum provided the temporal patterning of each of the warnings was distinctive, and yet good discriminability could be maintained. With navigation warnings, for example, it has been the tradition to differentiate between warning signals by varying the number of blows of the hooter. This system was clearly developed for pragmatic reasons, but psychologically the

Table 5.9 Sources of confusion in auditory warnings (from Edworthy, 1994b)

Source of confusion	Remarks
Continuous tone	Frequency difference irrelevant; sounds highly aversive and difficult to localise
Similar temporal pattern	Timbral differences largely irrelevant
Repeating on/off pattern	More generalised than sharing a temporal pattern; caused by similar proportion of on/off time
Similar start	Remainder of sound irrelevant, even if quite different
Similar labelling	Warnings quite different in many ways can be confused if they are labelled in the same way: e.g. 'repetitious', 'complex'

principle is quite ergonomic – especially if different and distinctive temporal patterns are used. Where only a different number of blows of the hooter is used there is the risk of confusion due to miscounting.

Another very important aspect of a warning is pitch, and pitch change. In the set of warnings described by McIntyre and Stanford (1985) only one of the warnings possessed a pitch change, which would make it very distinctive within this particular set. However, if several warnings within a set possess pitch changes, they might become harder to discriminate from one another.

Generally, one should be very careful in drawing conclusions from laboratory experiments in the case of pitch (to an extent this is also true of temporal patterning, although the effects are not as pronounced). Designing two warnings on the basis of a difference in pitch alone would not be a good idea. Experimentation might show that such warnings are readily differentiated when they are heard together, but the results of such experimentation would depend upon the precise paradigm used. Stimuli are easily discriminated on the basis of pitch when they are heard in close temporal proximity (Deutsch, 1978), but our ability to retain absolute pitch decays rapidly over time. Experiments show that two continuous tones at about 1 kHz and 3 kHz (but not exactly – meaning that octave equivalence does not account for these results) are readily confused in an experimental paradigm which forces the listener to retain information about each of the warnings over fairly long periods of time (Meredith and Edworthy, 1994). On the other hand, changes in pitch are extremely useful for altering the urgency of a sound over short periods of time, as we shall see in section 5.2.3 sounds.

To some extent the potential for confusion depends upon the range of warnings used in the set. Giving the users one of two warnings which have clearly identifiable labels is also useful, because they can then use the label to identify the sound, and make the association between the warning sound and the situation which it is signalling. For example, it would probably do no harm to have one siren present (although this would have to be modified to reduce its adverse characteristics), if its dangerous association is known (Lazarus and Hoge, 1986) and is to be capitalised upon. Certainly within a set of Patterson-type sounds a siren would be distinctive, and aid the learning of the entire warning set. One or two of the Stanford-type sounds would also aid discriminability. Thus a final point relating to making a warning set distinctive would be to open up the range of sounds used – though not to the point where the warning set sounds ridiculous (it would be rejected by the end-users) or aversive. It is important also to test the warning set as a whole, and perhaps some procedure based on the standardised procedure for evaluating information symbols (Easterby and Zwaga, 1984) provides the best template for the development of such a procedure. It has recently been proposed that standardised procedures for developing and evaluating warning symbols (Easterby and Zwaga, 1984) might be modified for use in auditory warning design and development, and this procedure is currently being tested (Stanton and Edworthy, 1994; Edworthy and Stanton, 1995).

5.2.3 Auditory Monitoring

5.2.3.1 Introduction

In many ways our auditory system is more 'real time' than our visual system. Gaver (1989) states that 'sound exists in time and over space, vision exists in space and over time'. Put another way, sounds are transient, but not location specific (although sounds come from a localisable source), whereas visual stimuli are located in one place, and can be more permanent. This contrast between the senses is the main factor in determining whether a visual or an auditory warning will be used under specific circumstances. Auditory stimulation tells us about the current state of the world, regardless of where we are looking or what we are doing. Auditory signals can therefore be used to great effect in monitoring events where we may not be focusing our main attentional capacity.

We use auditory feedback, naturally, nearly all of the time. For example, we listen to the changing sound of a car engine when we change gear; we dread the sound of a repeating 'clunk' from our disk drive, which we know means that our word processor has failed to read a data file or that there is some other problem with it; and we use auditory information all the time in determining the number, and location, of sound-producing objects in our environment. With the increasing capacity of computers to make all sorts of sounds, some efforts have gone into making artificial sounds for use in this type of auditory feedback. The use of warnings as icons has a great deal of relevance to the design of monitoring sounds, which can be thought of as quasi-warnings. They are icons, not of a specific situation, but of a situation which is changing over time. They provide more of the 'intelligent' or information part of a warning than more typical auditory warnings. Thus they can provide useful information for the hearer which can be more explicit than a warning itself. Much of the recent research on auditory warning design has direct application to the design of monitoring sounds.

Turning for a moment to the slightly more general idea of presenting information through sound, this has a number of attractions such as in applications for the visually impaired or as an additional cue in high-workload conditions. Examples of such applications are the use of auditory motifs in presenting chemical information to blind students (Lunney et al., 1983), the representation of graphs through sound (Mansur et al., 1985) and an auditory computer interface for blind users (Edwards, 1989). In these situations, auditory information is a replacement for visual information.

Status and monitoring sounds can act in a more informative way than we usually expect of warnings sounds, although their potential to interfere with the ongoing task is greater because of the necessary length of time these sounds must take. The sorts of sounds which give continuous feedback whenever a system is in operation – like the incessant beeping of cash registers, or the soundtrack of a video game – are too generalised to be considered here, but the use of sound in specific monitoring applications, and in the identification of problems which are impossible to detect visually, will be considered.

One of the central questions to be asked in the design of monitoring sounds is that of deciding which kinds of sounds to use. Inevitably, this involves some classification of sounds. One system of classification was proposed by Blattner *et al.* (1989), who distinguish between three types of auditory icon: the representational, the semi-abstract, and the abstract. This same kind of categorisation system is used for non-verbal visual signals, as discussed in Chapter 3. These can be image-related, i.e. representational, concept-related, or completely arbitrary. Most pictograms are somewhere between representational and completely arbitrary and are therefore the nonverbal visual equivalent of semi-abstract sounds. A further subdivision of the representational category has been proposed – icons can be divided into nomic, symbolic, and metaphoric representational icons. This system is shown in Table 5.10. Although such taxonomies are useful, there is little agreement as to which kinds of sounds are best suited for particular applications. There may, however, be some advantages in using some categories over others, which we will come to later.

The principles upon which classes of sound are mapped onto specific functions has been much speculated upon, but still relatively little explored in the research literature. For example, if we wish to create a monitoring sound for a pilot in order to convey to him or her that the engine is over-powering, the obvious solution might be to use the sound of the engine overpowering, even if the sound has to be resynthesised and sent through headphones. This view, and the whole concept of using everyday environmental sounds to convey trends, has been strongly and convincingly advocated by Gaver (1989). He has developed the concept of 'everyday listening', pointing out that when we hear a sound our first reaction is not normally to analyse its pitch, rhythm, or timbre, but to identify the source of that sound. Gaver cites several studies which show that when subjects are asked to produce free responses to a range of sounds, their overwhelming response is to name the object making that sound. Only when subjects are unable to do this do they describe the sound in terms of its physical characteristics. Gaver therefore makes a distinction

Table 5.10   Classes of sound

Abstract

Semi-abstract

	╱ nomic
Representational	─ symbolic
	╲ metaphoric

between everyday listening and musical listening, the latter being more focused on the physical attributes of sound. He advocates the use of environmental sounds in the design of an auditory interface for the Apple Mac computer (the SonicFinder), showing how sounds such as a jug filling can be used to mimic the action of a disk being copied, and the sound of an object being dragged as a mimic to dragging an icon across the desktop. He adds that using musical tones, or other more abstract auditory icons, has no link to the objects being represented and therefore their use is somewhat arbitrary. By using environmental sounds, and capitalising upon the nature of everyday listening, more intuitive systems can be developed. Such a system would therefore be of advantage in a number of situations in which monitoring sounds would be required.

However, there are a number of instances where environmental sounds would have a limited use in monitoring. Many operations have no clear environmental analogue so the selection of an appropriate environmental sound would be difficult and to some extent as arbitrary as assigning an abstract sound to that function. One would have to select some related environmental sound, hoping that the listener would both learn the relationship between the sound and the object or event which it is representing, and also forget the usual association with that sound. If an abstract sound is used, then one only has to ensure that the listener will make the required association – there is no other association which needs to be temporarily suppressed. Secondly, in some situations where the application of such sounds may be useful there might be a lot of background noise (such as in a helicopter) which prevents the use of such sounds. Here we may have some license to use only some of the elements of the sound. For example, there is a case for perhaps resynthesising and retaining only the important elements of the sound. Nosulenko (1990) has suggested that the resynthesis of real sounds does not necessarily involve a complete reproduction of these sounds, but depends more on the creation of sounds which cause the same images as the original sounds. The important question is which parts of a sound cause particular kinds of images? Fortunately there has been some work in this area, which is considered below.

For monitoring sounds, like warning sounds, it is important to establish empirical evidence of the relationship between acoustic parameters and psychological affect, or meaning. An early study was carried out by Solomon (1959a, 1959b, 1959c). Solomon studied sonar signals and their interpretation by sonarmen. He found that the signals were interpreted along several dimensions. These dimensions were magnitude, aesthetic value, clarity, security, relaxation, familiarity, and mood (Solomon, 1959a). In a second study (1959b), he attempted to isolate the acoustic parameters responsible for these interpretations. He found that the distribution of energy across frequency bands was important, correlating with nearly all of the meaning dimensions. He also found that the beat pattern was an important factor in the judging similarity between sounds, a result mirroring the finding of Patterson and Milroy (1980).

In a subsequent study (Solomon, 1959c), Solomon recreated sonar signals in which he systematically varied the levels of individual acoustic parameters and asked people other than sonarmen to judge the resulting stimuli. He found that the interpretations of these signals were similar to those found for sonarmen using real sonar sounds.

Another useful study which is relevant to the area of monitoring sounds is one by Warren and Verbrugge (1984). In a study of 'bouncing' and 'breaking' events, they looked for 'transformational invariants' – those features of sounds which convey what is happening to an object at any point in time. They differentiate between transformational invariants and 'structural invariants', which are usually constant and help in the identification of the object itself. After analysing the stages of breaking and bouncing, the experimenters looked at the acoustic consequences of these stages – that is, the acoustic changes taking place as a result of the event itself – and then extracted the transformational invariants specific to each type of event. The results of their experiments showed that the temporal pattern of the event itself gave enough information to allow the listener to distinguish between events. Warren and Verbrugge (1984) suggested also that spectral features are more important in the identification of objects, rather than in the identification of what is happening to those objects.

Taken together, the results of Warren and Verbrugge (1984) and Solomon (1959a,b,c) suggest that sounds can be resynthesised provided that we carry out good empirical studies on the acoustic parameters important in conveying particular meanings. This has application in the design of monitoring sounds where it may be necessary to convey several meanings at once. If we could isolate the specific associations between acoustic parameters and semantic meaning this could be used as the basis for the design of monitoring sounds. A project in which this was done is described in the following pages.

5.2.4   An example of the design and development of monitoring sounds

In this section, a practical example of the design of trend monitoring sounds is described, based on work carried out over the last few years (Loxley, 1991; Edworthy et al., 1992; Edworthy et al., 1995). In this project, five trend monitoring sounds were designed for use in helicopters. In order to design the sounds, empirical work on the meaning of sound was first carried out. On the basis of this work the monitoring sounds themselves were designed and the resultant sounds tested for confusability. The sounds were designed for use in helicopters, and the purpose of these sounds (named 'trendsons') was to indicate to the pilot that the acceptable limit of some flying parameter had been exceeded, but it had not yet reached a sufficiently dangerous level for a warning to be required. They were thus intended for use between normal and critical parameter values. In practical terms, the trendsons needed to perform two important functions. The first was to convey the meaning, as far as possible, of the situation being signalled (although pilots might reasonably be

Table 5.11   Pitch and speed stimuli tested by Edworthy *et al.* (1995)

Burst label	Number of pulses	Pitch sequence (Hz)	Interpulse interval sequence (ms)
Pitch level 1	6	100 × 6	0 × 5
Pitch level 2	6	250 × 6	0 × 5
Pitch level 3	6	400 × 6	0 × 5
Pitch level 4	6	550 × 6	0 × 5
Pitch level 5	6	700 × 6	0 × 5
Pitch level 6	6	850 × 6	0 × 5
Pitch level 7	6	1000 × 6	0 × 5
Speed level 1	3	300 × 6	950 × 2
Speed level 2	4	300 × 6	560 × 3
Speed level 3	5	300 × 6	375 × 4
Speed level 4	7	300 × 6	182 × 6
Speed level 5	9	300 × 6	85 × 8
Speed level 6	11	300 × 6	30 × 10
Speed level 7	12	300 × 6	0 × 11

expected to learn the sound-meaning association). The second was to convey increases or decreases in the parameter being monitored through the sound itself, and possibly also to match this with appropriate changes in urgency.

This work was carried out in several stages. The first stage was a free-association experiment in which three classes of sounds were described by subjects. The three classes of sounds used were from three categories. Class 1 was sounds produced by real objects (such as cars, corks popping and so on), Class 2 was music, which is completely abstract, and Class 3 involved semi-abstract sounds. The responses obtained were classified into groups, and the results showed that environmental sounds were overwhelmingly described in terms of their source, as Gaver's (1989) theory would predict. The results also showed that musical sounds were described largely in terms of their affect or mood. Semi-abstract sounds were most usually described in terms such as 'something is about to happen' and so on – a very useful feature for trendson design (Loxley, 1991). As semi-abstract sounds can be synthesised and therefore tailored to the situations for which they are required more readily than musical or environmental sounds, it was decided that small, semi-abstract units of sounds would be used as the basis for the trendson design. Many of the situations in which these sounds would be used range from just above normal to critical in a few seconds, suggesting that the basic building block of these sounds should be minimal. Furthermore, in the types of monitoring task for which the sounds were being designed – a helicopter – limits have to be defined, beyond which it is necessary to take action. For example, we know that the rotors of a helicopter can keep on slowing down until they stop, but we also know that the pilot needs to do something before that final state is reached. A situation where artificial limits are placed upon it requires sounds which can more directly mirror these restrictions (by design manipulation) and can be made to change rapidly with only small changes in the physical parameter they are monitoring. Environmental sounds, which have great potential in the design of monitoring sounds more generally, often change rather too slowly and imperceptibly for such a specialist application.

In the first stage of the study a list of 42 adjectives found to be relevant in some way (either through responses obtained to different types of sound, or by talking to pilots) to the design of monitoring sounds was derived. Next, a set of 28 very simple auditory stimuli were designed. These represented 7 levels each of rhythm, pitch, inharmonicity (whereby the number of harmonic components not bearing a simple integer relationship to the fundamental increased from Level 1 through to Level 7) and speed, which had all been found from earlier studies to be important in the design of auditory warnings. The stimuli were presented as short bursts of sound. Details of the pulse-by-pulse progression of these stimuli are shown in Tables 5.11 and 5.12. The building blocks of these bursts, the pulses, were all 200 ms in length with a 20 ms onset and offset envelope. A range of pulses was used across the stimuli depending upon the precise nature of the parameter under investigation. For example, the pulses in the inharmonicity set were each different across the seven levels because this was the parameter under investigation and therefore had to be changed. For some parameters there was no need for the pulses in each of the seven stimuli to differ, so the same pulse was used throughout. One example of such a parameter was speed, where the pulse-to-pulse interval rather than the spectral content was of interest.

All 28 stimuli were rated on a 1-7 Likert-type scale for the 42 adjectives selected. An 'irrelevant' box was available for each assessment. Unipolar scales were used because it is not always true that if a sound is not strong on one

Table 5.12   Harmonicity and rhythm stimuli tested by Edworthy et al. (1995)

Burst label	Number of pulses	Pitch sequence (Hz)	Interpulse interval sequence (ms)
Inhar level 1	6	300 × 6	0 × 5
Inhar level 2	6	300 × 6	0 × 5
Inhar level 3	6	300 × 6	0 × 5
Inhar level 4	6	300 × 6	0 × 5
Inhar level 5	6	300 × 6	0 × 5
Inhar level 6	6	300 × 6	0 × 5
Inhar level 7	6	300 × 6	0 × 5
Rhyth level 1	6	300 × 6	350, 275, 200, 125, 50
Rhyth level 2	6	300 × 6	300, 250, 200, 150, 100
Rhyth level 3	6	300 × 6	250, 225, 200, 175, 150
Rhyth level 4	6	300 × 6	200, 200, 200, 200, 200
Rhyth level 5	6	300 × 6	150, 175, 200, 225, 250
Rhyth level 6	6	300 × 6	100, 150, 200, 250, 300
Rhyth level 7	6	300 × 6	50, 125, 200, 275, 350

meaning it is necessarily strong on the opposite meaning – in many cases it is just irrelevant. The higher the rating, the more relevant was that particular stimulus on that particular meaning. The experiment was run in four sessions because of the fatiguing nature of the task, although in each session the stimuli were spread across the four parameters.

If, for a particular parameter-adjective pair, the overall response was 'irrelevant' in more than 15 per cent of the occasions where a rating could occur, that pair was excluded from analysis. The application of this heuristic left a total of 15 adjectives associated with the pitch stimuli, 13 with the speed stimuli, 12 with inharmonicity and only 5 with rhythm. The data were analysed for consistency across subjects (Kendall's W', which measures the extent to which rating consistency is achieved for individual stimuli regardless of the stimuli they are being compared with), and those parameter-adjective ratings resulting in significant levels of consistency were regressed. Those parameter-adjective mappings which resulted both in significant levels of consistency and significant regression coefficients (significantly linear trends) are shown in Figures 5.8 (pitch), 5.9 (speed), 5.10 (inharmonicity) and 5.11 (rhythm). These data show strong and significant positive mappings between pitch and 'rising', 'dangerous', 'urgent', 'high' and 'straining', and there are significant negative mappings between pitch and 'low' and 'safe'. Thus for the first five adjectives the strength of the adjective increases as pitch increases, and for the last two the strength of the meaning drops as the level of pitch rises (Figure 5.8). For the speed stimuli (Figure 5.9), three positive mappings were obtained, between speed and 'dangerous', 'urgent', and 'fast'. For inharmonicity (Figure 5.10), two mappings were obtained, both negative (meaning that as inharmonicity increased the ratings of those adjectives decreased). One of the mappings was between inharmonicity and 'powerful', and the other was between inharmonicity and 'urgent'. The rhythm stimuli were divided into two sets for analysis, because half of the stimuli speeded up throughout their duration (stimuli 1 to 4, with 4 being completely regular) and half slowed down

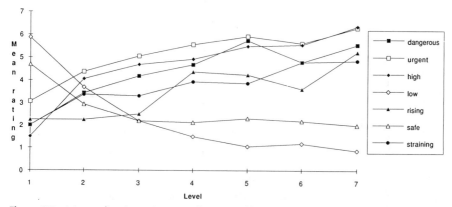

Figure 5.8 Mean adjective ratings as a function of level for seven pitch stimuli (from Edworthy et al., 1995).

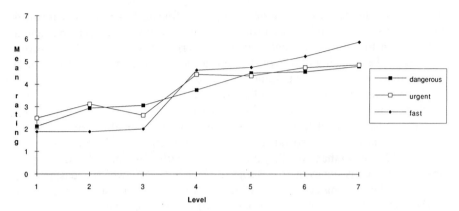

Figure 5.9 Mean adjective ratings as a function of level for seven speed stimuli (from Edworthy *et al.*, 1995).

(stimuli 4 to 7, again with 4 being regular). Only one significant mapping was found here, between the slowing down stimuli and the adjective 'jerky' (Figure 5.11). Many other associations were found between the acoustic parameters and the adjectives, but only the strongest ones are shown here. A complete account of these mappings can be found elsewhere (Edworthy and Hellier, 1992b). It is worth noting that the adjective 'urgent' shows a significant association for three of the four parameters tested, indicating once again the importance of this construct in both warning and monitoring sound design.

On the basis of these parameter-adjective mappings a set of five trend monitoring sounds was designed. These were for rotor overspeed, rotor underspeed, torque (power), positive g forces and negative g forces. Each of these presents a slightly different challenge in terms of design. Of the five, rotor

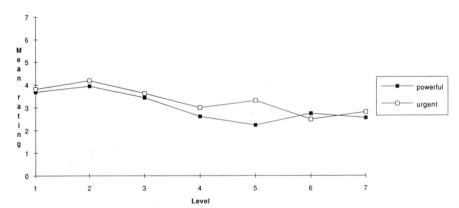

Figure 5.10 Mean adjective ratings as a function of level for seven inharmonicity stimuli (from Edworthy *et al.*, 1995).

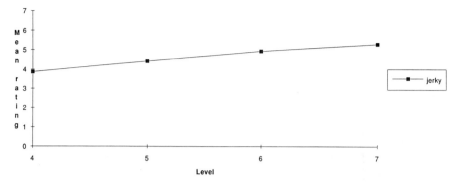

Figure 5.11 Mean adjective ratings as a function of level for four rhythm stimuli (from Edworthy *et al.,* 1995).

overspeed is the most straightforward because there are obvious acoustic images, such as speeding up and increasing in pitch. These changes also indicate appropriate changes in urgency. For example, an increase in speed is a useful analogue of the increasing speed of the rotors, and also indicates an increase in urgency. Rotor underspeed is a more difficult problem because acoustic analogues (such as slowing down of speed, or dropping of pitch), will also indicate a decrease in urgency which is of course in direct contrast to the urgency of the situation. Torque, or power, is reasonably straightforward because it is possible to create direct auditory analogues, and these are again likely to indicate appropriate changes in urgency – as the helicopter over-powers, so the situation becomes more urgent. The sounds for the g force trends are in some ways the most interesting, because there are no obvious acoustic analogues; however, things happen to the pilot's body during periods of both negative and positive g, and so it might be possible to produce a rather more abstract matching in this case. The trendsons for rotor overspeed, rotor underspeed and positive g are described below.

As well as the issue of which class or type of sound to use, the question of how the sound should be structured over time is important in the overall design of these sounds. One of the main questions is whether the sound – which could last for only a few seconds, or perhaps 5 minutes – should consist of a single, continuous sound or several discrete but related units. On the basis of previous empirical work (Edworthy *et al.,* 1992) a discrete structure with 5 levels was established as the protocol design for these trend monitoring sounds. This is shown in Figure 5.12. According to this protocol, separate but acoustically related sounds should be designed for each of the five levels of the trendson. This protocol, rather than one in which a continuously varying sound would be used, was chosen for a number of reasons. First, it is important for the pilot to know which of the stages he or she has reached, and in particular it is important for the pilot to know when a warning is imminent – this is the point at which critical values of the parameter under consideration have been reached, and the pilot should therefore not stay at this level for

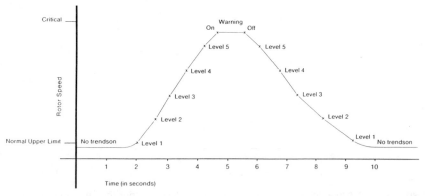

Figure 5.12 Protocol design for a trendson (from Edworthy *et al.*, 1992).

more than a few seconds. Provided the pilot is able to identify Level 5, the top level of the trendson, then a sound made up of discrete sounds is more likely to give the information needed to identify this level than is a unitary, continuously changing sound. A second reason for this type of design is that greater judgements of differences between small changes are obtained when separate, discrete units of sound are used (Edworthy *et al.*, 1992). This finding links into a great deal of scaling literature and, particularly in pitch, which is a crucial and very influential parameter under these conditions, it relates to the issue of prothetic and metathetic continua, one of the issues discussed at length by Stevens and Galanter (1957). This theory will not be discussed here, but is an important issue in perception which may have applications in other areas of warning design. For the purposes described here, the fact that pitch is a metathetic continuum – one in which increases or decreases in pitch produce qualitative, rather than quantitative, changes – encourages the use of units which change by steps rather than a continuously evolving sound without such steps. A third reason for selecting a five-stage protocol rather than a continuously changing one is that many situations where these sounds would be used switch from normal to critical in just a few seconds, and again a continuously varying sound might not convey the rapidity of such change as clearly as one with discrete steps. In practice, the set of five sounds constituting a trendson would be programmed into the analogue channels of the parameter being monitored and would be set to come .on once a particular value of that parameter had been reached. Each of the sound units was designed to be very small (only 200 or 300 milliseconds in length), and would play continuously until the next level is reached. If Level 5 is exceeded a warning sound is heard. If the level of the parameter drops below Level 1, then the trendson ceases.

In the design of the actual units of sound, one of the most interesting challenges was to create a set of five sounds that are related, but different in some way. A further challenge was to make Level 5 of the trendson appropriately distinctive, yet at the same time maintaining its membership with the rest of the group of individual sounds. In many ways this is similar to the process of composing a theme and set of variations, and depends very much on applying some aspects of the cognitive psychology of musical theme recognition (for example, Chew et al., 1982; Welker, 1982; Carterette et al., 1986). The question of the calibration of identity is central here. The constraints are tighter than for music because the nature of the trend has also to be conveyed through these sounds. Testing the similarities and differences both within and across trendsons will be discussed later, but first let us look at the design of the sounds themselves.

Details of the rotor overspeed trendson are shown in Table 5.13. Each burst of sound is described as possessing six pulses, although in practice the basic unit of design is a two-pulse burst repeated three times (in practice the two-pulse unit would also repeat for as long as the physical parameter was locked at that particular level). It can be seen that as each level of the trendson increases so does its pitch. The interpulse interval shortens, so that the complete burst sounds faster. It can also be seen that for all levels except Level 5, the rhythm is rather syncopated. Level 5 was deliberately made completely regular to convey distinctiveness at this level. The pulse itself was kept constant throughout, and was 100 ms in length with a 20 ms onset and offset envelope. It possessed 8 harmonics, all integer multiples of the fundamental. Thus the pulse had a distinctive pitch. The indicating parameters, those that vary between Levels 1 and 5, are pitch and speed, plus there is also a rhythm manipulation. The results obtained from the semantic meaning experiments showed that all the parameter-adjective mappings are consistent with changes in the trend. Increases in pitch convey increases in the meanings dangerous, urgent, high, rising and straining, as well as decreases in safe and low. All these mappings are consistent with an increase in rotor speed, which becomes an increasingly dangerous situation as rotor speed increases. The same is true of the speed-adjective mappings. Significant speed-adjective mappings were obtained for 'dangerous', 'urgent', and 'fast', all of which again are consistent with the trend. One could argue that one does not need to carry out experiments to design such an obvious trendson – for the design described above is indeed the most obvious to anyone who spends some time thinking about it – but it is useful and encouraging to show that such designs can be supported by empirical data.

Table 5.13 Structure of 'rotor overspeed' trendson (from Edworthy and Hellier, 1992b)

Trendson level	Number of pulses	Pitch sequence (Hz)	Interpulse interval sequence (ms)
1	6	400 × 6	100, 200, 100, 200, 100
2	6	420 × 6	75, 175, 75, 175, 75
3	6	440 × 6	50, 150, 50, 150, 50
4	6	470 × 6	0, 100, 0, 100, 0
5	6	500 × 6	0, 0, 0, 0, 0

The third indicating parameter – rhythm – was used to confer unique identity on the fifth level. Because a smaller number of mappings was obtained for
rhythm, it became clear that rhythm can be used as a relatively 'uncontami-
nated' acoustic parameter which can be used to great effect in the identification of the sounds themselves. Previous studies on auditory warnings
(Patterson and Milroy, 1980; Meredith and Edworthy, 1994) have shown that
temporal pattern (rhythm) is a very important source of warning identification
and confusion, and there are relatively small numbers of meaning attached to
rhythmic variations – at least this is true for the ones tested in the studies
presented here. The trendson increases in speed from Level 1 through to Level
5, but maintains a syncopated rhythm until Level 4. Between Levels 4 and 5
the rhythm changes from syncopated to regular, which makes Level 5 distinctive and thus identifiable. Of course, the problem here is that Level 5 might
sound too different from the others to be reliably identified as part of that
group of sounds. This issue was investigated in a further set of studies, to be
described later.

The design of a rotor underspeed trendson is more challenging. When this
situation arises the rotors themselves are slowing down to dangerously slow
speeds, which will ultimately become critical. There are clear acoustic analogues, and the immediate design response might be to design a trendson that
is in direct contrast to the rotor overspeed trendson; that is, one which both
slows down and drops in pitch. However, there are a number of reasons why
this might not be the best approach in terms of design. First, some of the
parameter-adjective mappings are directly inconsistent with the meanings that
should be associated with the trend itself. For example, dropping in pitch will
lead to lower associations with meaningful adjectives such as 'urgent', 'dangerous', 'safe' and 'straining', which is undesirable. However, there will be some
consistent mappings between the more directly analogous adjectives such as
rising, high and low. These mappings would be consistent with the trend, as
the rotor speed is indeed becoming low, and dropping. Thus there might be
some merit in dropping the pitch. Drops in speed would also produce inconsistent mappings on the more abstract adjectives ('dangerous' and 'urgent') and
produce one useful and consistent mapping for 'fast'. Thus, dropping one of
either pitch or speed might be viable, but certainly not both.

A second reason for not simply designing a trendson which is the exact
mirror image of the rotor overspeed trendson is that of confusion. Even
though the resulting sounds may differ spectrally, confusion could well occur
between a rotor overspeed trendson changing from Levels 5 through to 1 and
a rotor underspeed trendson changing from Levels 1 to 5, because each would
be signified by a dropping pitch and a slowing speed. Similarly, there would
also be confusions when the trendsons were heard in the reverse directions.
The sounds would have to be considerably different on other counts to avoid
confusion, which would present significant design problems.

On the basis of these two paradoxes, two different trendsons were designed.
One increased in speed but dropped in pitch, the other increased in pitch but
dropped in speed. Which of the two is likely to be more effective can only be

determined by the pilots themselves in field testing and in use. It depends to a great extent upon which are the more important images and associations for the pilots themselves, and requires further development work.

The third trendson to be described here is the one for positive g forces. In this case there is no direct analogue for the designer to use as a basis. On the other hand, increases in g forces create a sense of heaviness and pulling on the pilot, whereas decreases in gravitational pull (negative g) produce a sense of lightness and light-headedness. Although the most stringent data analysis (significant consistency across subjects and significant linear regressions) produced no parameter-adjective mappings which could be directly used for this application, a large of number of other, slightly weaker, associations were found. The adjectives 'full', 'heavy' and 'solid' produced associations with the inharmonicity stimuli, and so a trendson for positive g was designed where the number of irregular harmonics was increased through the five levels. The speed was also slowed down through these five levels. The associations between 'full', 'heavy' and 'solid' were only shown within the inharmonicity stimuli, making this parameter particularly useful in trendson design and application. Turning to the more robust mappings, the only mappings shown for the inharmonicity stimuli were 'powerful' and 'urgent' which were neither contradictory nor consistent with the trendson being conveyed. Among the less robust associations for speed, some were found between the adjectives 'straining', 'heavy' and 'tight' and slowing speed, all of which are also consistent with the trend.

Thus the design process was something of a compromise between the obvious analogues of the trend being conveyed – the reduction as far as possible of contradictory meanings, but the inevitability of having to accept some of them – and the judicious use of less robust associations between the acoustic parameters used and the adjectives tested.

The final part of the research programme was to test for confusions both within and between the trendsons. This was investigated via another project (Loxley, 1991). As well as conveying the appropriate meaning, there are two other important features of the five trendsons designed. The first is that each level of each individual trendson should be identified as belonging to that particular group (to a greater extent than it is perceived to belong to any other group of sounds) and the second is that each of the five levels of individual trendsons should to some extent be identifiable in an absolute way, especially the fifth level. The two are opposing requirements – to make five sounds appear related they need to be similar, but to make each level distinctive they each have to be different in some way – and thus another calibration issue comes to the fore. The individual levels of the trendsons were tested for these features, and the results of this work led to minor modifications of the sounds. First the extent of confusion between each of the individual levels of each of the trendsons was derived. In the next section, a study examining two of the trendsons – rotor overspeed and torque (power) – from the point of view of confusions, will be described.

Each subject was taught two of the trendsons, and given the name and the level of each in turn. Subjects were then required to name both the trendson and the level represented by that particular stimulus. An example of a confusion matrix obtained for power and rotor overspeed can be seen in Table 5.14. This shows that, out of 20 possible hits for each stimulus (an individual level of a trendson), subjects correctly identified the individual level of rotor overspeed 15/20 times for Level 1, 15/20 times for Level 2, 11/20 times for Level 3, 14/20 times for Level 4 and 20/20 times for Level 5. A similar, but not quite as good, pattern is shown for power. In both cases, the largest number of misidentifications occurred for the middle levels of the trendson, as one would expect. In power, these levels were mistaken for other levels of the same trendson, but in rotor overspeed there were a few occasions where the trendson itself was misidentified. However, identification of absolute levels of a trendson (a difficult task) was generally very good, and particularly so for the end-points, especially Level 5. Some combinations produced higher levels of confusion, and this was particularly noticeable for the rotor overspeed/rotor underspeed combination, which suggested that design modifications needed to be made.

In the second part of this investigation, subjects were required to rate the similarities between the individual levels of the trendsons using a 9-point scale. The trendsons were presented in pairs and the results for the rotor overspeed/power combination are shown in Table 5.15. These results show some interesting patterns, and demonstrate that the gradation of similarity had, to a

Table 5.14 Confusion matrix for rotor overspeed (RO) and Power (Po) trendsons. The rows show the trendson and level presented, the columns show the response given (from Loxley, 1991)

Stimulus Presented	Stimulus reponse										
	Rotor overspeed					Power					
	1	2	3	4	5	1	2	3	4	5	Total
Rotor overspeed											
1	15	5	0	0	0	0	0	0	0	0	20
2	2	15	3	0	0	0	0	0	0	0	20
3	0	5	11	3	0	1	0	0	0	0	20
4	0	1	2	14	2	1	0	0	0	0	20
5	0	0	0	0	20	0	0	0	0	0	20
Power											
1	0	1	0	0	0	15	4	0	0	0	20
2	0	0	0	1	0	5	12	2	0	0	20
3	0	0	2	0	0	1	3	8	5	1	20
4	0	0	2	1	1	0	1	7	8	0	20
5	0	0	0	1	2	0	0	1	2	14	20
Total	17	27	20	20	25	23	20	18	15	15	

Table 5.15 Mean similarity ratings between trendson levels of rotor overspeed (RO) and Power (Po). Pairs of stimuli were rated on a scale from 1 (very dissimilar) to 9 (identical) (from Loxley, 1991)

	Rotor overspeed					Power				
	1	2	3	4	5	1	2	3	4	5
Rotor overspeed										
1	9.0	2.8	2.6	2.2	2.4	1.2	1.6	1.2	1.4	1.4
2	2.8	9.0	6.8	3.4	3.0	1.2	1.4	1.2	1.6	1.4
3	2.6	6.8	9.0	3.8	2.4	1.2	1.4	1.4	1.6	1.4
4	2.2	3.4	3.8	8.4	2.6	1.4	1.4	2.0	1.6	1.2
5	2.4	3.0	2.4	2.6	9.0	1.4	1.6	1.2	1.6	1.4
Power										
1	1.2	1.2	1.2	1.4	1.4	9.0	5.0	5.6	3.6	3.4
2	1.6	1.4	1.4	1.4	1.6	5.0	9.0	6.4	4.4	3.2
3	1.2	1.2	1.4	2.0	1.2	5.6	6.4	9.0	5.0	5.6
4	1.4	1.6	1.6	1.6	1.6	3.6	4.4	5.0	9.0	6.2
5	1.4	1.4	1.4	1.2	1.4	3.4	3.2	5.6	6.2	9.0

greater extent, been achieved in the design process. For example, the similarity between different levels of a trendson always decreased as the number of steps between them increased. Thus, for example, Level 3 of the rotor overspeed trendson was more similar to Levels 2 and 4 than it was to Levels 1 and 5. Another interesting feature of this data is that the similarities between trendsons was generally much lower than those within trendsons. For example, the only similarity between rotor overspeed and power which was rated above a 3 on the 9-point scale was that between Level 4 of rotor overspeed and Level 4 of power. An intriguing adjunct to this was that often the similarities across trendsons depended upon level. For example, the similarity pattern within the Level 3 of the rotor overspeed trendson is mirrored, at a much lower level, in the power trendson. That is, Level 3 of the power trendson rated as more similar to Level 3 of the rotor overspeed trendson than were any of the other levels of the power trendson. The level of similarity dropped as the number of steps between levels increased (so Levels 2 and 4 of power were rated as being more similar to Level 3 of the rotor overspeed trendson than were Levels 1 and 5). This suggests that there was also something fairly distinctive about the way the level of each trendson was conveyed, which was common across trendsons. A more significant point is that the difference in similarity ratings between Level 1 and its closest step up, and between Level 5 and its closest step down, was larger than for any other stepwise changes. This confirms that these levels were more distinctive than others, which was part of the design remit. Of course, if these levels are too different from one another then confusions will occur, but the confusion matrices (Table 5.14) show that the confusion level was fairly low. The fact that Levels 1 and 5 were correctly identified more often than the other levels (and were not mistaken for other trendsons) suggests that the calibration of sameness and differentness was about right in this design task.

The main purpose of these studies was to check on the similarity calibration both within and between trendsons. The research largely confirmed that the design was about right, but there were one or two cases where individual stimuli were either too similar to or too different from one another, and on the basis of these findings they were altered slightly, retested for confusion, and the specification of this set of trendsons finalised. They were then tailored to the acoustic environment using software based on Patterson's guidelines, and at the time of writing are in the process of being flight tested.

CHAPTER SIX

Speech warnings

6.1 Introduction

A short story by the author John Irving entitled 'Interior Space' (Irving, 1993) tells of the domestic life of a young urologist and the trials and tribulations of his patients. The hospital in which he works has an ingenious system for dealing with cardiac arrest. 'Dr Heart' is summoned over the intercom in a very calm voice and asked to go to the patient's room. The rule is that any doctor should discretely get him or herself to that room to attend to the problem urgently. So much better, points out the author, than yelling at the top of one's voice in panic, with the statutory number of 'Oh my God!'s in the response that one has come to expect from any disaster movie worth its salt. The advantage of such a system is that patients and visitors are not startled or worried, yet the staff working in the hospital are able to decode the meaning of the message with ease, and they can then act on that information appropriately. In practice, the meaning of the warning was only too well understood – the general rule was that doctors would amble slowly to the nearest lift in the hope that some other doctor would get to the patient first.

Calm, almost perfunctory, synthetic or digitised computer voices are the stuff of science fiction films. These disembodied voices, conveying the most important and urgent of information in a style so laconic that they might be reading out names and addresses from a telephone directory, are memorable aspects of at least two classic science fiction movies, *2001: A Space Odyssey* (we must all remember 'Hal') and *The Andromeda Strain*. In the latter, a major nuclear incident is mere seconds away, but the computerised voice tells the scientists and support staff in the unit under threat what to do to avoid a catastrophe. Unfortunately, the nonverbal warnings sounding at the same time are not subject to such sophistication. The interesting thing about these computerised voices is the intonation chosen by their human imitator. At the time these films were made the available speech technology was not capable of providing the signals that were used (indeed it was the use of such voices that was being predicted by these films) yet it was almost taken for granted that when these systems were to be used to warn of imminent danger they would do so in a calm, matter-of-fact manner.

At the simplest level, the information parts of a spoken message are the words themselves, and the iconic parts are the prosodic, paralinguistic elements of that message. When we talk naturally, the two usually go together and our voices are flat, without any particular intensity. When we are angry, it shows not just in the words that we use but also in the speed, pitch and intensity of the signals themselves. Furthermore, as speakers as well as listeners we are able to dissociate the iconic and the information at will. We know that cutting and rude remarks often have a greater impact if we deliver them in a nonstressed voice and we can normally tell when a speaker does not actually mean what he or she is saying by the prosodic elements of the message. So when we are listening to speech we take heed of both linguistic and paralinguistic elements, which may either be matched, or mismatched and inappropriate. Urgency mapping might therefore be of some consequence in the development of speech warnings, although at present there is very little research in this area. There are one or two research projects which have studied this issue, however, and we will discuss them later.

Language perception and production is a uniquely human skill, and is highly developed. We might expect, therefore, that warnings using language would be the most effective of all warning types. Spoken warnings add an extra dimension to both written and nonverbal auditory warnings. In written warnings they add the paralinguistic elements absent from written warnings (the iconic features of the design can substitute for these paralinguistic elements) and in the case of nonverbal warnings they add the linguistic element (nonverbal auditory warnings can be very strong on iconic, paralinguistic elements). It remains to be seen whether this is true, however. There are still many technological problems which need to be overcome if, for example, spoken warnings are to be as resilient to masking as synthetic nonverbal auditory warnings; there are problems of intelligibility and detectability when spoken messages are used in environments where other speech communication is used; and there are methodological issues in cross-modal warnings research which need to be addressed before the apparent suitability of speech as a warning medium can be confirmed. Speech warnings have been directly compared with nonverbal tones (Simpson and Williams, 1980; Hakkinen and Williges, 1984), with written warnings (Wogalter and Young, 1991; Wogalter et al., 1993) and with pictorials (Byblow, 1990), and in general they have come out favourably. However, we need to examine other factors which may contribute to these effects as well as the simple modality differences, which we will also discuss later.

Speech warnings have been advocated for both noisy and 'quiet' environments. In the Sections 6.2 and 6.3 we will consider research which has looked at these two environments. In noisy environments, intelligibility is a central issue so this will be looked at briefly. A lengthy description of intelligibility indices and acoustics is not carried out, as detailed information is available elsewhere (see for example chapters by Sorkin (1987) and Simpson et al., 1987). Our main focus will be the exploration of the likely effectiveness of spoken warnings, and how they have, and may be, compared with warnings in other modalities. The issue of perceived urgency will be explored in some detail.

6.2 Voice warnings in quiet environments

Spoken warnings can in some circumstances be used as a substitute for written warnings, and there is some research which has compared warnings in various modalities, including speech. There have generally been favourable effects for spoken warnings, although some ambiguity exists. This research has been largely reviewed in earlier chapters, but there are some further points which can be drawn from it which are directly relevant to a discussion of speech warnings. One is that spoken warnings might be usable in open environments, those where people are moving at a high rate. These are the kind of environments where we would normally expect visual warnings to be used. In Chapter 2, an experiment by Wogalter and Young (1991) was discussed which involved a chemical mixing task, but included an experimental manipulation where a voice warning, given by the experimenter, was tested as well as a print condition and a print-plus-voice condition. There was no control condition, so compliance scores only are given. The experiment showed a significant difference between the voice-and-print and the print-only conditions, but there were no other significant effects. In a second experiment, the experimenter did not speak the warning, but a tape recorder with the spoken message was used in its place. The experimenter started the tape, but the participant stopped it. Unfortunately no voice-only condition was included in this experiment. The results showed a significant difference between the voice-and-print and the print-only conditions, in favour of the former. This enhanced effect of voice warnings on compliance was also demonstrated in a subsequent field study. A slippery floor hazard was set up in a shopping precinct, and mops, cones and so on were set up to make the hazard look realistic. People walking by were warned in one of four conditions: no warning (just the cones, bucket and mop), a print-only warning, a voice-only warning (on a tape), and a print-and-voice condition. Thus in the field study the voice-only condition was reinstated and the control condition was included. When no warning was present 20 per cent of the people that walked by exhibited behaviour that was classified as compliance. In the print-only condition 42 per cent of the people that walked by complied (giving an effectiveness score of 22 per cent). In the voice-only condition 64 per cent complied (effectiveness score 44 per cent) and in the voice-plus-print condition 76 per cent complied (effectiveness score 56 per cent). These results confirm the advantage for spoken warnings and demonstrate that composite warnings appear to be more effective than either of the modalities alone, although not quite additively so.

There are some issues of calibration and comparison which need to be discussed with respect to these findings. From an ergonomist's point of view we may not be too concerned as to which particular features of a warning lead to its increased effectiveness – we may simply wish to implement those designs which are found to be most effective. As empiricists, however, we might wish to delve deeper into the possible causes and investigate the alternative variables which might be having some effect in these studies. To what extent, for example, have the results been due to the fact that potentially important features of stimuli have not been equalised across experiments? To what extent has consideration been given to the more complex, possibly esoteric, issues which relate to the arguments about equalisation of perceived urgency across variables which we have discussed at length in this book? We turn now to some of these issues.

It is important to disentangle the social influence factor from the modality issue, as Wogalter and Young attempted to do by replacing the experimenter's voice (and presence) by a tape. However, it is also important that the intensity level at which different spoken warnings are presented is kept at a fixed level – it is not clear in this study whether the taped message was presented at the same intensity level as the message delivered by the experimenter. We know that the level at which an nonverbal auditory warning is presented has a large impact on its perceived urgency (for example, Momtahan 1990, Haas and Casali, 1995) so it would be important to control for this in future experimentation. Another important variable which we know affects compliance rates is temporal and spatial proximity. Spatial proximity is not really an issue with auditorily-presented warnings – provided the warnings are audible this variable is unlikely to have much effect on compliance. Temporal proximity, however, may well have an effect, so it would be important to ensure that spoken warnings in different forms are presented at the same moment in the sequence of the experiment. It is fairly clear that this was true for the live and taped warnings in Wogalter and Young's (1991) experiment, and it was also true in the experiment where the written warning was presented at the same time as the speech warning. It is important to preserve this temporal equality in future experiments.

Controlling for temporal proximity across modalities is much more complex, and raises some interesting questions. In Wogalter and Young's (1991) field study the spoken warning was presented on a tape loop every 10 seconds, but the warning sign was continuously visible (providing, of course, that the passer-by was looking at it). For some passers-by the spoken warning would have been playing at precisely the moment they walked past the hazard, and in some it would not have been. In the case of the visual warning, it should have been visible at the point of passing for all subjects. To assist in determining precise causes, several refinements could be incorporated in future related studies. One is a spoken warning which is initiated when the participant crosses an infrared beam (this was incorporated in the Wogalter et al. 1993 study, discussed at length in Chapter 2) so that everybody hears the warning at the same distance from the obstruction. Another is a flashing or illuminated visual warning, with the same words as the spoken warning and a duty cycle identical to that of the spoken warning. Yet another would be to construct visual and spoken versions of the same warning and to ensure that each is presented at the same distance from the obstruction, and the same amount of time elapses between first sight (or hearing) of the warning and the eventual obstruction.

One experiment which provides a very simple and straightforward cross-modal demonstration is that carried out by Letourneau *et al.* (1986) who took simple reaction measures in response to a green light. Participants were required to turn off the light as soon as it came on by pressing a button. In each trial, the onset of the green light was preceded by a buzzer or a white light, both of which were presented at 1 metre from the participant. The results show that reaction times were faster to the green light when an auditory, rather than a visual, warning was given. This experiment also explored the effects of alertness on reaction times, which we will not discuss here. In experiments of this kind, however, it may be possible to control the sensory intensity of the visual and the auditory warning signal by carrying out pre-experimental matching studies of a psychophysical nature. If some agreement could be reached as to what level of light intensity was approximately equivalent to a particular sound intensity (one needs to do rather more than present these stimuli from equal distances if true comparability is to be achieved), then the two types of warning could be directly compared to provide a more accurate picture of any advantage the auditory modality may have over the visual, or vice versa, in producing simple reactions to simple physical stimuli.

Of course, one of the reasons for using an auditory warning of any kind is that it can be used in instances where a visual warning may not be noticed, because the person to whom the warning is addressed is not looking in the right direction. This could certainly be true in a busy shopping area, and the results of Wogalter and Young's (1991) experiment shows the added advantage of using voice warnings in such circumstances. However, we believe that it is important to demonstrate these effects as unambiguously as possible in order to build up a testable model of compliance and inform the calibration issue, as we suggested in Chapter 1.

Before moving on, let us turn briefly to the most abstract and complex calibration issue, that of cross-modal calibration and equalisation. In Chapter 1 we proposed several methods of achieving this, settling on the 'ergonomic range' argument. Research carried out on such a basis would allow us to build up a database of the relative effects of different variables on compliance rates, effectiveness scores and so on. We might then be able to predict the size, colour and/or wording that would be required of a warning sign so as to make its apparent urgency the same as that of a specific spoken warning. The same is true for the spoken warning – we might be able to suggest ways in which the prosodic and informational elements should be used to produce warnings of specific urgencies, for example. Thus, we might in future be able to say that the apparent urgency of a warning sign of a specific (apparent) size, colour and general design was approximately equally to that of a spoken message with a specific content, intonation pattern and intensity. Warnings matched in this way could then be used in cross-modal comparisons. The results would give clearer insights into any advantages that specific modalities might have in influencing warning compliance.

Let us conclude this section by proposing that we should not take cross-modal comparisons at face value, but should perhaps decide for ourselves (depending upon our particular focus of interest) the level at which we are prepared to consider warnings in different modalities to be equivalent. There are surprisingly few studies which compare spoken warnings with written ones, and there is certainly much scope for research work in this area in the future. What little research there is suggests that voice warnings can be quite effective. The Wogalter et al. (1993) study discussed in Chapter 2, for example, showed much greater effects for a voice warning than for any of the other types of warning presentation, of which there were many. Not all studies have demonstrated this advantage, however.

When warnings in two modalities are presented almost simultaneously, compliance generally increases. Our intuition might predict this, and the dual-coding theory (Paivio, 1985) provides theoretical support. Paivio has argued that cognitive processes involve two independent processes, the imaginal and the verbal. When the two are operating together, memory is facilitated. Thus, presenting warnings simultaneously in different modes may improve memory for that warning, which in turn may increase compliance rates. The potential for increasing compliance rates therefore looks promising, but we have to consider the consequences of a wholesale application of this factor. First of all, if, as argued earlier in this book, it is the role of the warning to reflect the level of risk rather than simply to induce compliance regardless of any other circumstances, then adding more modalities in a particular instance may ultimately reduce the receiver's future predisposition towards complying with warnings. Additionally, bombarding people with a warning in a number of modalities may be appropriate in a relatively warning-free environment, but we may need to think carefully before doing so in an environment where many warnings are present, either simultaneously or in rapid temporal succession, such as in any process-plant control room.

A few studies where composite warnings have been considered, almost incidentally, have been those which have measured the compliance of subjects who were previously shown videos of warnings-related behaviour (Racicot and Wogalter, 1992; Chy-Dejoras, 1992). In a study by Racicot and Wogalter (1992) participants were shown either a silent video of a warning sign, a video shot of the warning sign followed by a video shot of protective equipment, followed by shots of someone putting on the protective equipment (again silent), or the modelling video with a male voice announcing the warning. The results showed higher compliance rates for both videos with the modelling, but no additional effect for the voice-over (probably a ceiling effect, as compliance scores were 92 per cent in the silent modelling condition, and 100 per cent in the voice-over condition).

Advertising campaigns are another area of practical interest. Since they are often initiated by the manufacturer of a potentially hazardous product it is only compliance scores which are of interest in practice. The manufacturer may not be concerned about calibration, but is simply concerned that his or her warnings are complied with as often as possible. As advertising campaigns may be conducted in a written mode (magazines, newspapers and so on) or a written and verbal mode (TV advertisements), the relative compliance scores produced by these media is of some practical consequence. The dual-modality research to which we have already referred would predict that TV advertising would be more effective, because it is using both visual and auditory modalities. Alcohol warnings have been of some interest in the US in recent years, and this has produced a number of research papers in that area (Cvetkovitch and Earle, 1995; Hilton, 1993). Barlow and Wogalter (1993) looked specifically at magazine and television warnings, but found rather ambiguous results. The dual-modality effect is largely confirmed, as would be predicted by theory. Barlow and Wogalter (1993) point out that some studies which have directly compared print and spoken warnings have shown a superiority for print warnings, whereas others have shown a superiority for spoken warnings. Under the most controlled conditions, however, it is our opinion that the evidence lies in favour of spoken warnings.

6.3   Voice warnings in noisy environments

If spoken warnings are to be used in noisy environments (which they frequently are), there are a number of complex technical issues, most of which are to do with intelligibility, which need to be addressed. Although we will not discuss these in detail here, we will go over some of the basic issues before considering studies which have looked at spoken warnings in noisy environments. In most cases spoken warnings have been compared with warnings in other formats or modalities. There seems, however, to be a dearth of studies which have compared spoken and visual (verbal) warnings in noisy environments.

Assuming that a spoken warning is to be generated in some way by a piece of equipment (rather than produced by a live voice or by a recording of the entire warning), then there are two main ways in which the warning can be produced. Messages can be entirely synthesised, that is, constructed artificially from rules without any humans being involved, or they can be constructed from real human speech which has been digitised and stored and then assembled as required. Where the messages are entirely synthesised the computer which produces the resultant messages contains rules and information about basic units of sound (such as phonemes) and acoustic parameters, voicing, and so on, from which the text is generated. Where the starting point is real human speech, the human speech is digitised, compressed in some way (using Fourier transform information, for example) and then recorded. The required speech is then recreated as needed from the recorded speech units.

In both cases, even in quiet surroundings, there are some intelligibility problems, especially with synthesised speech, although these problems are being reduced as speech systems improve in their sophistication. If voice signals are required for a noisy environment, and normal quiet speech is amplified, then if the consonants are adjusted to an appropriate loudness level the vowels will be too loud. If the vowels are set at an appropriate level the consonants may not be heard. This can become a more acute problem if some of the higher frequencies (the higher formants) of the consonants are filtered out by the system or, because of the complexity of the noise environment, are rendered inaudible, thus causing further confusion as to the identity of particular consonants. Of course, language processing is a top-down, as well as a bottom-up, process so a listener is often able to fill in the missing information. But under conditions of high workload, redundancy in speech information becomes more important, as we shall see later.

Another problem associated with both methods of message production (more so for synthesised speech than for digitised speech) is that of providing appropriate pitch and stress patterns for the resultant speech messages. We will also come to these later in a slightly different context.

Cowley and Jones (1992) provide a useful comparison of synthesised and digitised speech, providing a checklist which can be used to inform the decision as to whether a digitised or a synthesised voice message is likely to be the more suitable mode for any particular application. Their comparison is shown in Table 6.1. Some of the more pertinent points (in terms of the focus we are taking in this book) of this table warrant a little further discussion. If the setting is noisy, then Cowley and Jones recommend the use of synthesised rather than digitised speech. This is mostly because it is easier to match a wholly artificial acoustic signal to a complex noise spectrum than it is to boost acoustic signals which come from a human speaker (in much the same way that it is possible to create a synthetic nonverbal auditory warning for a complex noise environment, as described in detail in Chapter 5). Another point worthy of note is that Cowley and Jones recommend the use of digitised, rather than synthesised, speech if the operator is carrying out some other arduous task at the same time. The main reason for this is that the processing of synthesised speech seems to impose greater demands on cognitive processes than does natural speech (the reason may have something to do with the relative lack of pitch and stress patterns in synthesised speech), meaning that there are fewer resources left for other tasks. This was demonstrated by Luce et al. (1983) who showed that performance decrements in a digit recall task fell off more rapidly with synthesised, as opposed to natural, speech as task complexity increased. However, technological advances have been great since that study was carried out so the difference may not now be as noticeable. Two further points of interest are that Cowley and Jones (1992) recommend a digi-

Table 6.1 Choice criteria (synthesised or digitised) for speech devices (from Cowley and Jones, 1992)

Message creation	Digitised	Synthesised
Setting is noisy	no	yes
User is an accomplished speaker	yes	no
User has advanced technical skills	no	yes
Message requires editing	no	yes
Computer memory is restricted	no	yes
Message has untypical pronunciations	yes	no
Message is confidential	no	yes
Speed of creation essential	yes	no
Task is arduous (but not spatial)	yes	no
Prosodic features essential for meaning	yes	no
Complex message (may require transcription)	no	yes
Message reception		
Speaker must be identified	yes	no
Noisy environment	no	yes
Receiver is overloaded with information	yes	no
Visual display or print-out necessary	no	yes
Message is lengthy	yes	no
Message carries warning/alerting function	no	yes

tised voice if prosody is important, but they also suggest that a synthesised voice may be better if the message is intended as a warning. The implication is that a warning message does not need prosody to be effective – the disembodied, attention-getting qualities of synthesised speech may make it more appropriate for use in warnings. This brings us back to the issue of the dissociation between the iconic and the informational elements of spoken messages, which is how we began this chapter. It is a topic to which we shall return later.

Intelligibility and recognisability are essential components of any artificial speech production system, and it goes without saying that, if speech warnings are to be used, the designer should ensure that intelligibility is optimised. There are many source documents (such as Simpson *et al.*, 1987) to assist the designer in maximising intelligibility. Intelligibility of a warning, however, does not guarantee compliance. Relatively few studies have been carried out on speech warnings, and of those, few have examined compliance, but before discussing the issue of compliance we shall examine some of the relevant studies on their own terms.

6.4 Tones used with voice warnings

Those studies that have been carried out generally compare speech warning systems with other warning methods, such as auditory tones or pictorials, using reaction time to the warnings as the main outcome measure. As with nonverbal auditory warnings, research has been more concerned with the technological problems of producing the signals than with eventual compliance, which is only to be expected given the complexity of those technical problems. Often the main motivation for the study is a comparison between an older system using, say, a nonverbal warning such as a bell or a tone, and a warning generated by the new technology of digitised or synthesised speech. We should point out that the 'older' nonverbal warnings used in such comparisons seldom involve best practice such as described in the present volume.

Experiments which have compared various combinations of voice messages and tones have raised numerous issues concerning the relative merits of the two types of warnings, and of their combination. One issue is whether the presence of an alerting tone increases reaction time or not; a second is whether a simple tone can serve an alerting function, even though it may impart no information of itself; a third issue is the effect of informational redundancy on responses to verbal warnings; and a fourth is the issue of whether different word presentation rates affects intelligibility or other factors connected to the perception of such warnings.

A study by Simpson and Williams (1980) demonstrates the effect of using an initial tone as a simple alerting signal when it precedes a warning message given in a noisy environment. In this experiment five voice warnings were tested in two formats. The two formats were a 'keyword' format, where only the minimum number of words (two or three) was used to convey the situation to the participant. The other format was a 'semantic context' condition where a fuller linguistic context was provided by the voice message, which in all cases increased the length of the message by one word (the additional word providing more specific context for the hearer). The two sets of stimuli are shown in Table 6.2. Each of these conditions was tested in two different contexts; in one the words were preceded by a tone, and in the other they were not. As the tone was undifferentiated across the five verbal warnings, and was a very simple auditory cue, it could not have given the hearer any additional information. At

Table 6.2 Warning voice formats (semantic context and keyword) and durations (from Simpson and Williams, 1980)

Semantic context	Duration (s)	Keyword	Duration (s)
Landing gear not down	1.78	Gear not down	1.25
Tank boost pumps out	1.67	Boost pumps out	1.32
Cabin pressure dropping	1.77	Pressure dropping	1.31
Collision traffic one o'clock	1.88	Traffic one o'clock	1.36
Engine fuel filter bypass	1.87	Fuel filter bypass	1.40
Flight instruments disagree	1.88	Instruments disagree	1.48
Mean message duration (s)	1.80		1.35
Standard deviation	0.080		0.070

most it could serve as a rather low-level alerting signal, indicating 'a warning is following', so it serves in a somewhat different capacity to that of the auditory warnings discussed in Chapters 4 and 5. Results indicated that adding the tone increased the overall reaction time by 0.65 seconds, which was less than the fixed one-second time interval between the beginning of the tone and the beginning of the following voice. Thus the tone did indeed serve in some alerting capacity. With regard to the different voice warnings, the additional semantic context was found not to affect reaction times, even though we might have expected the longer message to have taken longer to assimilate. We will look at the semantic context effects in a little more detail in a few moments.

In a similar paradigm, though with a different, and quieter, noise background – the task was an air traffic control simulation task, so only fairly low-level office noise was used – Hakkinen and Williges (1984) showed that the use of similar alerting tones (and also alerting lights) increased reaction time. However, in some of the conditions it was found that when a synthesised voice message was preceded by an alerting tone more of the messages were detected. So in this experiment a simple auditory tone, along with a light, had no positive impact on one important dependent variable, reaction time, but was shown to have a positive impact on a yet more crucial dependent variable, detectability. This demonstrates an important feature of even the most basic of auditory tones which makes them useful in many environments, namely that they can be made appropriately loud in such a way as to be detectable in difficult and complex noise environments, as we have discussed at length in earlier chapters.

It is important to note that these two experiments did not explore the use of other, richer, nonverbal auditory tones. It is not surprising that a signal with neither any informational value (other than signifying that a warning was coming) nor iconic value (the tones used were the same throughout) simply served to slow down reaction times. In the task used by Simpson and Williams (1980) and Hakkinen and Williges (1984), participants were required to listen to the voice warning before acting, and so were required, at least to some extent, to assimilate the semantic content of the message before reacting. However, one more recent experiment (James and James, 1989, discussed in Chapter 2) shows that richer auditory warnings with some urgency mapping and prioritisation can produce faster reaction times. The question of how auditory warnings which are designed as a replacement for verbal warnings (because they are suitably prioritised and learned beforehand) perform relative to matched spoken warnings is a topic which has received no research attention other than the brief experiment by James and James (1989), and would seem to be a useful topic for future research.

A research topic which stems from these issues is the potential effectiveness of nonverbal auditory warnings which produce readily-learned mnemonics, or can have words fitted to them. In terms of design there are no major problems of setting such a warning at appropriate loudness levels, and if the word-matching of these signals is easily learned, or intuitive, then this might prove valuable in situations where the noise environment is simply too complex for adequate voice warnings. For example, the cardiovascular alarm described in Chapter 4 (Figure 4.2), has six pulses and is readily heard as a mimic of the word 'cardiovascular', and so can function as a 'nonverbal' verbal warning (or a 'verbal' nonverbal warning). The efficacy of such designs is yet to be tested, but this is another direction which might prove fruitful in warning design and application. An offshoot of this design suggestion is that such warnings might function in a similar way to keyword type verbal warnings, because there are representations of only short (usually single word) phrases, similar to the 'keyword' warnings in Simpson and Williams' (1980) study. The research literature suggests that such keywords are not as effective, in terms of reducing reaction time, as more linguistically rich warnings, to which we now turn.

6.5 The semantic content of voice warnings

Another finding from Simpson and Williams's (1980) study was that the more semantically rich format (the three- or four- word warnings, rather than the two- or three-word keyword format) warnings did not produce slower reaction times, even though the semantically richer warnings were an average of 0.3 seconds longer than those in the keyword format. This replicates an earlier finding (Simpson and Hart, 1977). In fact, responses to the keyword format were slower than they were to the semantically richer warnings, although not significantly so. We can join the authors in interpreting this as showing that greater redundancy of information in the warning reduces processing time per word, which also decreases the amount of attention required in processing the message. Thus semantically richer voice warnings might be particularly useful under high workload conditions.

The effect of linguistic redundancy on reaction time has a number of interesting parallels in written warning design. Consider the fact that the 'semantic context' warnings were generally more explicit than the 'keyword' warnings (Table 6.2). For example, the first warning states 'landing gear not down' in the semantic context format, and 'gear not down' (leaving the recipient to work out which particular gear is not down) in the keyword format. Laughery et al., 1993a, (see Chapter 2) showed that for written warnings the explicitness of the warning increased the perceived hazardousness of the product to which the label was referring. It is just possible that the same thing happens for vocally presented warnings, and that the increased perceived hazardousness of the situation results in a reaction which, on the basis of warning length alone, may have been slower but has been speeded up because of the greater perceived hazardousness resulting from the increased explicitness of the warnings. This is of course highly speculative, but the dissociation of warning length, redundancy and explicitness might be a fruitful area for future research, perhaps not in the sphere of ergonomics, but from a more theoretical perspective.

A study which confirms the complexity of the issues at stake here is one by Byblow and Corlett (1989) who also looked at the effects of linguistic redundancy on reaction time. In this study, participants performed a flight simulation task during which warnings were presented in six different formats – a warning in keyword format preceded by a tone, a warning in semantic format preceded by a tone, two warning formats consisting simply of keywords or words in semantic formats (no tone), and two warnings in which the tone was effectively replaced by a coded word 'Danger!', 'Caution!', or 'Warning!' In both the tone and coded word condition the combination of tone/word plus silence added an extra 0.5 seconds to the complete warning. The results showed that the keyword condition produced the fastest reaction time, and there were no significant differences between any of the other conditions. However, the authors qualify their results by adding that in the keyword condition the reaction time measured from the end (not the start) of the warning was longest in the keyword format. This suggests that participants took rather longer to work out what to do in the keyword conditions than in some of the other conditions, and suggests also that the attention demands of this condition were the highest. The attention demand issue was also discussed in this way by Simpson and Williams (1980), who found that more semantically rich warnings did not increase reaction time, but may have made the task easier to perform.

Another parallel between the redundancy issue in spoken and written warnings is found between the ratings studies of Wogalter et al. (1987) (discussed at length in Chapter 2) and the Simpson and Williams study. Simpson and Williams (1980) suggest that, as the longer warnings appeared to be more informative, each one could be preceded by an appropriately urgency-coded word such as 'Danger' or 'Caution', as was subsequently tested by Byblow and Corlett (1989), although they did not differentiate and calibrate these keywords in the way that has been done in warning label research. Wogalter et al. (1987) showed that the signal words are the most redundant pieces of information in a written warning, so according to one view this redundancy should serve to shorten reaction time even more. On the other hand, these words are iconic, rather than informational, so they provide a different sort of information for the recipient than the words contained in the 'semantic context' warnings. Much the same information, in fact, as that which could be provided by an appropriately coded, prioritised, nonverbal auditory warning. There has been very little research into urgency coding and mapping of spoken warnings, even though it has been explored in some depth for both written and nonverbal warnings. We will turn to this later in the chapter, but before we do this we need to look at some of the other issues relevant to spoken warnings.

Two more recent experiments consider speech interpretation and intelligibility from a slightly different viewpoint, and these begin to suggest some questions relevant to the perceived urgency issue, which we shall come to at the end of the chapter. An experiment by Simpson and Marchionda-Frost (1984) looked at the effect of speech rate and pitch effects on warning intelligibility, and a study by Slowiaczek and Nusbaum (1985) takes a more general perspective on speech intelligibility. Both of these studies suggest that both the rate and pitch of speech has an impact on the intelligibility of the message.

Simpson and Marchionda-Frost (1984) presented pilots with threat warnings, which they were asked to report as they participated in a complex simulated flying task. The speech warnings were presented at three different fundamental frequencies (70, 90 and 120 Hz fundamentals) and three different speech rates (123, 156 and 178 words per minute), in a fully factorial design. Reaction times, measured from the start of the onset of the voice warning to the point where the pilot began to take action, showed a significant effect for speech rate, but no similar effect for pitch. Reaction times were fastest for those warnings presented at the fastest rate. Pilots also rated the warnings on semantic differential scales, from which it could be concluded that they marginally preferred the warnings presented at 156 words per minute, fearing that they might miss messages presented at a higher rate. Generally, the warnings presented at the middle rate were rated higher on those factors which might be important in the practical implementation of voice warning systems. However, if we look at the reaction time data alone then we would conclude that the highest rate – 178 words per minute – was the best because it produced the fastest reaction time without any apparent loss in intelligibility. Whether this is caused because more words can be conveyed in a shorter period of time or because the faster presentation rate may result in greater perceived urgency (or, indeed, whether the two are related in some way) remains an issue for exploration.

The intelligibility issue was explored in further detail by Slowiaczek and Nusbaum (1985). In the second of their two experiments participants were asked to transcribe synthesised voice messages which were presented at a rate of either 150 or 250 words per minute. In addition, the messages were presented either with a flat pitch contour (as a monotone) or with a natural pitch contour superimposed upon them. The dependent variable of interest here was not reaction time, but the number of words correctly transcribed, thus the experiment was not about speech warnings *per se*, but about intelligibility more generally. The results are shown in Figure 6.1. This shows that the messages presented at the slower rate were more accurately recognised than those presented at the faster rate. It also shows that those messages in the 'short' format (sentences were presented in two different lengths, 'short' and 'long') produced higher percent correct rates than those in the 'long' format, which might be expected. There was an effect of the type of sentence (one of three types – active, passive, or embedded), and there was a smaller, but significant, effect for pitch contour. The nature of this effect was that the messages presented in the 'inflected' condition (with a pitch contour imposed on them, similar to that produced by natural speech) produced higher scores than those presented in the 'monotone' condition. There were some interactions, but these appear to be mainly due to the degree of linguistic complexity in the message.

Thus both Simpson and Marchionda-Frost (1984), and Slowiaczek and

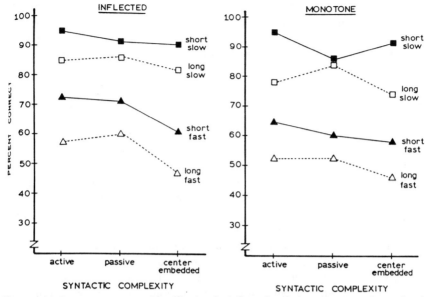

Figure 6.1   Percentage correct identification for inflected and monotone speech signals. The filled symbols correspond to short sentences and open to long. The squares represented the sentences presented at a low rate (150 wpm) and the triangles a fast rate (250 wpm) (from Slowiaczek and Nusbaum, 1985).

Nusbaum (1985) show that pitch contour and speech rate have some effects on the intelligibility of speech to the extent that very fast rates reduce intelligibility. The introduction of a pitch contour seems to produce a small improvement on the intelligibility of speech but does not appear to be as important a factor as some of the other factors relevant to the design of speech messages. There is room, however, for further disentanglement of the effects of relevant variables. In speech rate in particular there is scope for more detailed calibration of the loss of intelligibility as speech rate increases. For example, although Slowiaczek and Nusbaum (1985) showed that the higher speech rate (a very high rate of 250 words per minute) affected intelligibility, Simpson and Marchionda-Frost (1984) found no objective impairment in performance at their highest speech rate, which was 178 words per minute. What they did find was a preference for the middle speech rate, which might be anticipated on the grounds of aesthetics. The reported preference for the middle rate in Simpson and Marchionda-Frost's study was that words might be missed at the faster rate, but there is little objective evidence to support this fear. In future warnings work it will also be important to dissociate aesthetic preferences from actual performance measures, especially as Simpson and Marchionda-Frost's (1984) study shows that the fastest reaction times were achieved in response to the warnings presented at the fastest rate, suggesting that this is in fact the best rate.

Since the rate of a spoken message is clearly measurable and definable, studies where the relationship between intelligibility and rate are explored over a larger set of variable levels are achievable. Such experimentation would give data about the point at which intelligibility really does drop off. It may be useful, in perceived urgency terms, to know just how fast a warning can be presented with no loss of intelligibility. In many applications, of course, the rate of speech presentation is not crucial, but in the design of warnings this information might very well be crucial. Another topic which might provide data which can be applied is the point at which aesthetic judgement (or preference) and performance dissociate, especially if less preferable designs actually produce better performance.

Generally, spoken warnings have either been compared with one another in different formats, as the studies reported here show, or they have been compared with nonverbal auditory warnings. However, a study by Byblow (1990) directly compared speech displays with pictorial displays. Comparisons of different display types are fairly common fare in human factors, and in general they are concerned with cross-modal performance and integration, relying heavily on Wickens' multiple resources model (e.g. Wickens, 1992) for their theoretical underpinning. Such studies are generally not concerned only with the input modality, but also with the output modality, as the multiple resources model has much to say and predict about this too. The topic of multiple resources is too large and general for the purposes of this particular book, as is the general topic of display modality, but there are very many issues which need to be explored within a broader framework. However, we are including the study by Byblow, as the types of messages used in his study are spoken warnings, and are therefore directly relevant to the topic of this book. Also, they compare spoken and pictorial warnings, which is something of a rarity, and they demonstrate some effects of linguistic redundancy that are rather at odds with those of Simpson and Williams (1980).

An earlier study by Robinson and Eberts (1987) directly compared speech and pictorial displays in an experiment where participants were heavily engaged on a flight simulation task. From time to time they were presented with either speech or pictorial information and reaction times were measured. The results showed that participants responded more quickly in the pictorial condition. It was also found that the speech displays placed a greater strain on short-term memory, mostly because of the modality difference itself – in the pictorial condition participants were more able to 'zoom in' (the authors' words) on the pictorial display. This particular experiment demonstrates tight control of potentially confounding cross-modality factors as the displays were presented in fixed time slots. Thus the advantage of the pictorial display was not due to its permanence, as it was presented for exactly the same time period as the speech display.

In Byblow's (1990) experiment participants were presented with messages of high or low redundancy in either a graphic or a speech format. High redundancy messages were generally those which gave extra information which was not necessarily needed in order to decode the message. In addition, there was a warning-only condition where participants were required to carry out a simultaneous flight task as well as to respond to the warnings, and a dual-task condition in which participants were involved in a secondary task at the same time as the flight task and the warning task. Amongst other results there was no significant effect for modality (although reaction times were faster in the speech condition than in the pictorial condition) and there was a significant effect for redundancy. Here, it was found that the low-redundancy messages produced faster reaction times than high-redundancy messages, a finding which replicates the earlier finding of Byblow and Corlett (1989) discussed earlier, and one which does not fully support the earlier Simpson and Williams (1980) finding.

Thus the findings for the cross-modality comparison are to some extent equivocal. This is not surprising as the issues and interactions are so complex and numerous. In addition to the calibration issues we have been raising in relation to many of the arguments presented in this book (which become very complex when a visual, nonverbal modality is being compared with an auditory, verbal modality) consideration must be given to all of the other issues relating more directly to the multiple resources model (Wickens, 1992) such as S-C-R compatibility, which are likely to have effects on dependent variables such as reaction time and overall task performance, especially when complex tasks are being considered along with responses to warnings. Such issues are beyond the scope of this book. However, we propose that the calibration issue should, and could, be incorporated quite broadly into the multiple resources model. This would allow cross-modal comparisons to be made more directly, and it would give us some idea, when we are carrying out cross-modal studies, whether we are comparing 'like with like', a question we have often raised.

6.6 Perceived urgency

In a written warning the iconic aspects are generally readily dissociated from the informational. For example, one could easily have a warning label where the words themselves would lead one to believe that there is a high level of risk or hazard involved with a product even though the printed words are small, unclear and generally not attention getting. In spoken messages the informational and the iconic aspects can also be readily dissociated, as we pointed out at the beginning of this chapter, but this is more difficult, as in spoken warnings the two naturally go together. We have seen how, in synthesised speech, the removal of the natural prosodic, iconic features of the message reduces intelligibility and, to some extent, the believability of the message.

The appropriate and inappropriate association of the iconic and the informational content of verbal warnings generally may be important, but thus far there is little research evidence which can be brought to bear on this question.

The information part of a warning is that part which enables the recipient to make the decision as to whether or not to comply with a warning. The iconic part provokes some more basic response, hopefully one which is appropriate even if the information is not properly understood or interpreted. There is no particular reason why the same should not be true of spoken warnings, although the extent of the importance of the prosodic, or as we have interpreted them, iconic, aspects of spoken messages are as yet unknown. We might expect the perceived urgency of a spoken warning to be influenced by the pitch, speed and intensity of the message. These issues have been explored in nonverbal auditory warning design, but not in relation to spoken warnings, perhaps because the informational elements present in spoken warnings are so overwhelming in relation to the iconic aspects that the latter have been neglected.

The effect of speech rate and pitch have been explored in relation to message intelligibility, as discussed earlier, but studies of the perceived urgency of speech (whether in relation to warnings or otherwise) have yet to be done. Intelligibility studies have also looked at the effects of the complete absence of prosodic elements in synthesised speech, but again the perceived urgency issue has yet to be explored. Intuitively we suspect that the complete non-use of appropriate pitch contour, for example, has some effect on urgency, but there is no research evidence to suggest this. The atonal, repetitive pitch contours used by trained speakers whose job is to give out information in environments such as (most notably) aircraft provide anecdotal support for the view that pitch contour might have some importance in the perception of urgency. Airline stewardesses (who seem to be particularly adept at this skill) start off each new statement at a relatively high pitch, fall throughout the statement, with one or two rises along the way, until they reach a low pitch and the end of the statement. Everything they say follows this same pattern. We cannot thus differentiate between a bland statement about the purchasing of duty-free goods and the essential safety information that we also receive, at least in iconic terms. The reasons why the information is presented in this way might well be of interest to a variety of professionals, but we will not speculate upon the matter here.

Recent studies (Haas and Edworthy, 1995) have shown that a major proportion of the variance attributable to perceived urgency effects in nonverbal auditory warnings can be attributed to their pitch, pulse rate and intensity, and on the basis of this a perceived urgency model for auditory warnings is being developed. A topic of considerable interest would be to see if a similar model could be produced for speech perception. However, we would not expect the values appropriate for such a model to be the same – considerable pitch changes are required to produce changes in perceived urgency in nonverbal auditory warnings, but we would expect the pitch changes required to produce perceived urgency changes in spoken warnings to be much smaller, as the human voice pitch range is considerably smaller than that which is possible for nonverbal auditory warnings. On the other hand, it may be that because the pitch range of the human voice is not large the relative contribution of other parameters to a composite urgency model is larger than for non-verbal auditory warnings. This whole area is open to research impetus, and there is much that could be done, and much that could be speculated upon.

One highly relevant experiment is a study by Sorkin *et al.* (1988). Participants performed a tracking task (the primary task) and also a simultaneous monitoring task. The monitoring task was one in which, at pseudo-random intervals, a set of four 3-digit numbers appeared briefly at the bottom of the monitor being used for the tracking task. The value of each number was somewhere between 0 and 8, and the higher the numbers the more likely a signal was to be present. After each set of numbers had been presented, participants had to decide either to report that a signal had occurred, or to do nothing if they considered that a signal had not occurred because of the low value of the numbers. The task was therefore a signal detection task in which the likelihood of the signal varied from trial to trial.

In some of the conditions used by Sorkin *et al.* (1988) the likelihood of the signal was also represented by a visual or spoken warning, which also coded the level of likelihood of the signal. The visual and auditory warnings each occurred in two-state and four-state versions. The visual warnings were given by the colour of the cursor in the tracking task. The two-state visual warnings were either white or red, depending upon the urgency (likelihood) of a signal being present. The four-state visual warnings were either white, green, yellow or magenta, again depending upon the likelihood of a signal being present. The two-state spoken warnings were either no message or the message 'check signal', and the four-state spoken warnings were no message, 'possible signal', 'likely signal' and 'urgent signal'. The combinations tested in this experiment are shown in Table 6.3, where the numbers correspond to the total score of three numbers between 0 and 8 which the participant was expected to monitor, and the level at which each of the warnings was presented. The higher these numbers, the more likely a signal was to be present.

There are numerous questions of interest in this study. The type of alarm (colour or speech) had no overall impact on the monitoring task; the 4-state alarm produced better performance than the 2-state alarms in the difficult, but not the easy, tracking task; and there was an interaction between alarm state and alarm type whereby performance was better in the speech condition initially (in the no alarm and 2-alarm states), but performance continued to improve at a greater rate when the 4-state colour warnings were used. In addition, performance accuracy increased and reaction time decreased when alarm displays were used (as opposed to no alarms). Thus the presence of the alarm system generally improved performance, even though one might have expected it to increase the workload of the operator. Furthermore, the more complex alarm design generally produced the same increase in performance as the simple alarm display, although in the difficult tracking task performance on the monitoring task improved between the two alarm states.

This experiment shows clearly the effect that the presence of an alarm can

Table 6.3 2-state and 4-state visual and speech warnings, showing criterion score triggering alarm condition (from Sorkin et al., 1988)

Condition	Display	Alarm criterion
Visual 2-state		
	white	0.0
	red	3.5
Visual 4-state		
	white	0.0
	green	3.0
	yellow	3.5
	magenta	4.0
Speech 2-state		
	no message	0.0
	'check signal'	3.5
Speech 4-state		
	no message	0.0
	'possible signal'	3.0
	'likely signal'	3.5
	'urgent signal'	4.0

have on performance, but leaves some questions still to be explored, such as the relationship between workload and alarm complexity, and the possibility that other types of alarm design (even different types of spoken alarm) might have had different effects on performance. For example, the three annunciated levels of the 4-state verbal alarm were 'possible signal', 'likely signal' and 'urgent signal'. The three words used to describe the signal might not be differentially graded levels of the same dimension, with the consequence that there might be better calibrated words to use.

In Sorkin et al's (1988) study, urgency was represented by the likelihood that an alarm state actually existed. There is some recent evidence (Bliss et al., 1995) to suggest that people are very good at matching their alarm response rate to signal believability, so further investigations along this line would almost certainly prove to be fruitful.

6.7 Conclusion

Spoken warnings represent a concatenation of the research problems inherent in both warning label and nonverbal auditory warning design, and in this chapter we have drawn out some of these issues and suggested some areas for future research effort and related them, where possible, to already-studied topics in both of these areas. The topic of perceived urgency in spoken warnings is certainly worthy of research effort, and spoken warnings are ideal as experimental stimuli because they allow the dissociation of the iconic from the informational. Such delineations are also possible in warning labels and non-verbal auditory warnings, but spoken warnings may be the most natural tool for the exploration of such dissociation.

Spoken warnings have sometimes been compared with warnings in other modalities (often because they are replacing older, nonverbal or nonauditory, warnings). Such comparisons bring to the fore many of the issues involved in cross-modality comparisons which we have raised in this book.

References

ADAMS, A. S. and EDWORTHY, J. (1995) Quantifying and predicting the effects of basic text display parameters on the perceived urgency of warning labels: trade offs involving font size, border weight and colour. *Ergonomics*, **38**(11), 2221–37.

ADAMS, A. S. and MONTAGUE, M. (1994) Testing warning signs: Conspicuity and discrimination, *Information Design Journal*, 7, 203–10.

AIGA (1981) *Symbol Signs*, New York: American Institute of Graphic Arts.

AJZEN, I. and FISHBEIN, M. (1980) *Understanding Attitudes and Predicting Behaviour*. Englewood Cliffs, New Jersey: Prentice-Hall.

ANSI Z535 (1987) *Criteria for Safety Symbols*, American National Standards Institute.

AS 2342 (1992) *Development, Testing and Implementation of Information and Safety Symbols and Symbolic Signs*, Sydney: Standards Australia (PO Box 1055, Strathfield NSW 2135, Australia).

ASCH, S. E. (1955) Opinions and social pressure, *Scientific American*, **193**, 31–5.

AUSTRALIAN DEPARTMENT OF HEALTH, HOUSING and COMMUNITY SERVICES (1991) *National Survey of Chemicals Used in the Home and the Community's Understanding of the Hazards of Domestic Chemicals*, Phillip, Australian Capital Territory: Author.

AYRES, T. J., GROSS, M. M., HORST, D. P. and ROBINSON, J. N. (1992) A methodological taxonomy for warnings research, *Proceedings of the 36th Annual Meeting of the Human Factors Society*, 499–504. Santa Monica: Human Factors Society.

BAIRD, R. N., McDONALD, D., PITTMAN, R. K. and TURNBULL, A. T. (1993) *The Graphics of Communication*, 6th edn. Fort Worth: Harcourt Brace.

BANDURA, A. (1977) Self-efficacy: toward a unifying theory of behavioural change, *Psychological Review*, **84**, 191–215.

BLATTNER, M. M., SUMIKAWA, D. A. and GREENBERG, R. M. (1989) Earcons and icons: the structure and common design principles, *Human-Computer Interaction*, **4**, 11–44.

BLISS, C. K. (1968) *Semantography-Blissymbolics*, Sydney: Semantography-Blissymbolics Publications.

BLISS, J. P., GILSON, R. D. and DEATON, J. E. (1995) Human probability matching behaviour in response time to alarms of varying reliability. *Ergonomics*, **38**(11), 2300–12.

BOERSEMA, T. and ZWAGA, H. J. G. (1989) Selecting comprehensible warning symbols for swimming pool slides, *Proceedings of the Human Factors Society 33rd Annual Meeting*, 994–8, Santa Monica: Human Factors Society.

BOERSEMA, T., ZWAGA, H. J. G. and ADAMS, A. S. (1989) Conspicuity in realistic scenes: An eye-movement measure, *Applied Ergonomics*, **20**(4), 267–73.

BRAUN, C. C., SANSING, L., KENNEDY, R. S. and SILVER, N. C. (1994) Signal word and colour specifications for product warnings: an isoperformance application, *Proceedings of the 38th Annual Conference of the Human Factors and Ergonomics Society*, 1104–8. Santa Monica: Human Factors and Ergonomics Society.

BRESNAHAN, T. F. and BRYK, J. (1975) The hazard association values of accident-prevention signs, *Professional Safety*, **20**(1), 17–25.

BROWN, S. L. (1992) The hazard meter: a conceptual heuristic tool of risk assessment, in LAVE, L. B. (Ed.), *Risk Assessment and Management*, 499–507. New York: Plenum Publishing Co.

BURT, J. L., BARTOLOME, D. S., BURDETTE, D. W. and COMSTOCK, R. J. (1995) A psychophysiological evaluation of the perceived urgency of auditory warning signals, *Ergonomics*, **38**(11), 2327–40.

BYBLOW, W. D. (1990) Effects of redundancy in the comparison of speech and pictorial displays in the cockpit environment, *Applied Ergonomics*, **21**(2), 121–8.

BYBLOW, W. D. and CORLETT, J. T. (1989) Effects of linguistic redundancy and coded voice warnings on system response time, *Applied Ergonomics*, **20**(2), 105–8.

CAIRNEY, P. and SLESS, D. (1982) Communication effectiveness of symbolic safety signs with different user groups, *Applied Ergonomics*, **13**, 91–7.

CARTERETTE, E. C., KOHL, D. V. and PITT, M. A. (1986) Similarities among transformed melodies: the abstraction of invariants, *Music Perception*, **3**(4), 393–410.

CHAPANIS, A. (1994) Hazards associated with three signal words and four colours on warning signs. *Ergonomics*, **37**, 265–76.

CHEW, S. L., LARKEY, L. S., SOLI, S. D., BLOUNT, J. and JENKINS, J. J. (1982) The abstraction of musical ideas, *Memory and Cognition*, **10**(5), 413–23.

CHILLERY, J. A. and COLLISTER, J. B. (1988) Confusion experiments on auditory warning signals developed for the Sea King helicopter. Technical memo FS(F) 688, Royal Aerospace Establishment, Farnborough, UK.

CHILLERY, J. A. and COLLISTER, J. B. (1986) Possible confusions amongst a set of auditory warning signals developed for helicopters. Technical memo FS(F) 655, Royal Aircraft Establishment, Farnborough, UK.

CHY-DEJORAS, E. A. (1992) Effects of an aversive vicarious experience and modelling on perceived risk and self-protective behaviour, *Proceedings of the Human Factors Society 36th Annual Meeting*, 603–7. Santa Barbara: Human Factors Society.

COLE, B. L. and HUGHES, P. K. (1990) Drivers don't search: they just notice, in BROGAN D. (Ed.), *Visual Search*, pp. 407–17, London: Taylor & Francis.

COLE, B. L. and JENKINS, S. E. (1980) The nature and measurement of conspicuity, *Proceedings of the Australian Road Research Board Conference*, **10**, 99–107.

COLLINS, B. L. and PIERMAN, B. C. (1979) *Evaluation of Safety Symbols*, US

Department of Commerce, National Bureau of Standards, NBSIR 79-1760.

COLLINS, B. L. and LERNER, N. D. (1982) Assessment of fire-safety symbols, *Human Factors*, **24**, 75–84.

COOPER, J. B. and COUVILLON, L. A. (1983) Accidental breathing system disconnections. Interim report to the Food and Drug Administration. Cambridge, Massachussetts: Arthur D Little Inc.

COWLEY, C. K. and JONES, D. M. (1992) Synthesized or digitized? A guide to the use of computer speech, *Applied Ergonomics*, **23**(3), 172–6.

CUNITZ, R. J. (1981) Psychologicaly effective warnings, *Hazard Prevention*, May/June, 5–7.

CVETKOVITCH, G. and EARLE, T. C. (1995) Product warnings and information processing: the case of alcohol beverage labels, *European Review of Applied Psychology*, **45**(1), 17–20.

DEJOY, D. M. (1989) Consumer product warnings: review and analysis of effectiveness research, *Proceedings of the 33rd Annual Meeting of the Human Factors Society*, 936–9. Santa Monica: Human Factors Society.

DEJOY, D. M. (1991) A revised model of the warnings process derived from value-expectancy theory, *Proceedings of the 35th Annual Meeting of the Human Factors Society*, 1043–7. Santa Monica: Human Factors Society.

DEUTSCH, D. (1978) Delayed pitch comparisons and the principle of proximity, *Perception and Psychophysics*, **23**, 227–30.

DEWAR, R. E. and ELLS, J. G. (1977) The semantic differential as an index of traffic sign perception and comprehension. *Human Factors*, **19**, 183–9.

DEWAR, R. E., ELLS, J. G. and MUNDY, G. (1976) Reaction time as an index of traffic sign perception, *Human Factors*, **18**, 381–92.

DINGUS, T. A., WREGGIT, S. S. and HATHAWAY, J. A. (1993) Warning variables affecting personal protective equipment use, *Safety Science*, **16**(5/6), 655–74.

DOLL, T. J. and FOLDS, D. J. (1986) Auditory signals in military aircraft: ergonomics principles *vs.* practice, *Applied Ergonomics*, **17**, 257–64.

DREYFUSS, H. (1972) *Symbol Sourcebook*, New York: McGraw-Hill.

DUNLAP, G. L., GRANDA, R. E. and KUSTAS, M. S. (1986) Observer perceptions of implied hazard: safety signal words and colour words, Technical Report TR 00.3428. Poughkeepsie, New York: IBM.

EASTERBY, R. S. and HAKIEL, S. R. (1977) Safety labelling and consumer products: Field studies of sign recognition, *AP Report 76*, Applied Psychology Department, University of Aston in Birmingham.

EASTERBY, R. S. and ZWAGA, H. J. G. (1976) Evaluation of public information symbols ISO tests: 1975 series. *AP Report 60*, Applied Psychology Department, University of Aston in Birmingham.

EASTERBY, R. S. and ZWAGA, H. (1984) *Information Design*. Chichester, UK: Wiley & Sons.

EDWARDS, W. (1954) The theory of decision making. *Psychological Bulletin*, **51**, 380–417.

EDWARDS, A. D. N. (1989) Soundtrack: an auditory interface for blind users, *Human-Computer Interaction*, **4**, 45–66.

EDWORTHY, J. (1994) Urgency mapping in auditory warning signals, in STANTON, N. (Ed.), *Human Factors in Alarm Design*, 15–30. London: Taylor and Francis.

EDWORTHY, J., LOXLEY, S. L. and DENNIS, I. D. (1991) Improving auditory warning design: relationship between warning sound parameters and perceived urgency, *Human Factors*, **33**(2), 205–31.

EDWORTHY, J. and HELLIER, E. J. (1992a) Auditory warnings for the European fighter aircaft. Report on British Aerospace project No. 6012CA. London: British Aerospace.

EDWORTHY, J. and HELLIER, E. J. (1992b) Trend audio information systems. Report on MoD project No. 2021/14/EXR(F).

EDWORTHY, J. and STANTON, N. A. (1995) A user-centred approach to the design and evaluation of auditory warning signals. 1. Methodology. *Ergonomics*, **38**(11), 2262–80.

EDWORTHY, J., LOXLEY, S .L., GEELHOED, E. and DENNIS, I. D. (1988) An experimental investigation into the effects of spectral, temporal and musical parameters on the perceived urgency of auditory warnings. Report on Ministry of Defence project No. SlS42B/205. Unpublished manuscript, University of Plymouth, UK.

EDWORTHY, J., LOXLEY, S. L. and HELLIER, E. J. (1992) A preliminary investigation into the use of sound parameters in the portrayal of helicopter trend information. Report on Ministry of Defence contract No. SLS42B/568. Unpublished report, University of Plymouth.

EDWORTHY, J., HELLIER, E. J. and HARDS, R. A. J. (1995), The semantic associations of acoustic parameters commonly used in the design of auditory information and warning signals, *Ergonomics*, **38**(11), 2341-61.

ELIOT, W. C. (1960) Symbology on the highways of the world, *Traffic Engineering*, **31**, 18–26.

ELLS, J. G. and DEWAR, R. E. (1979) Rapid comprehension of verbal and symbolic traffic sign messages, *Human Factors*, **21**, 161–8.

EN 475 (in press) Medical Devices – electrically-generated alarm signals.

FINLEY, G. A. and COHEN, A. J. (1991) Perceived urgency and the anaesthetist: responses to common operating room alarms, *Canadian Journal of Anaesthesia*, **38**, 958–64.

FLESCH, R. (1948) A new readability yardstick, *Journal of Applied Psychology*, **32**, 221–3.

FMC CORPORATION (1985) *Product Safety Sign and Label System* (3rd ed), Santa Clara, CA: Author.

FOSTER, J. J. (1991) *Proposed Revised Method for Testing Public Information Symbols*, ISO TC145/SC1 Document N 220.

FOSTER, J. J. (1994) Evaluating the effectiveness of public information symbols, *Information Design Journal*, **7**, 183–202.

FRANTZ, J. P. and RHOADES, T. P. (1993) A task-analytic approach to the temporal and spatial placement of product warnings, *Human Factors*, **35**(4), 719–30.

FREED, D. and MARTENS, D. (1986) Deriving psychophysical relations for timbre. *Proceedings of the International Computer-Music Association*, 393–405.

GAVER, W. W. (1989) The SonicFinder: an interface that uses auditory icons, *Human-Computer Interaction*, **4**, 67–94.

GIBSON, G. (1979) *The Ecological Approach to Visual Perception*, New York: Houghton Mifflin.

GODFREY, S. S., LAUGHERY, K. R., YOUNG, S. L., VAUBEL, K. P., BRELSFORD, J. W., LAUGHERY, K. A. and HORN, E. (1991) The new alcohol labels: how noticeable are they? *Proceedings of the 35th Annual Meeting of the Human Factors Society*, 446–50. Santa Monica: Human Factors Society.

GODFREY, S. S., ROTHSTEIN, P. R. and LAUGHERY, K. R. (1985) Warnings: do they make a difference? *Proceedings of the 29th Annual Meeting of the Human Factors Society*, 669–73. Santa Monica: Human Factors Society.

GODFREY, S. S., ALLENDER, L., LAUGHERY, K. R. and SMITH, V. L. (1983) Warning messages: will the consumer bother to look? *Proceedings of the 27th Annual Meeting of the Human Factors Society*, 950–4. Santa Monica: Human Factors Society.

GOLDHABER, G. M. and DETURCK, M. A. (1989) A developmental analysis of warning signs: The case of familiarity and gender. *Proceedings of the Human Factors Society 33rd Annual Meeting*, 1019–23, Santa Monica: Human Factors Society.

GRAY, P. G. (1964) Drivers' understanding of road signs, *Traffic Engineering and Control*, **6**, 49–53.

GREEN, P. and PEW, R. W. (1978) Evaluating pictographic symbols: An automotive application, *Human Factors*, **29**, 103–14.

GRUMAN, G. (1994) New faces, *Australian PC World*, December/January 1995, 86–98.

GULLIKSEN, H. and TUCKER, L. R. (1961) A general procedure for obtaining paired comparisons from rank orders, *Psychometrika*, **26**(2), 173–83.

HAAS, E. and CASALI, J. G. (1995), Perceived urgency and response time to multi-tone and frequency-modulated warning signals in broadband noise, *Ergonomics*, **38**(11), 2313–26.

HAKKINEN, M. T. and WILLIGES, B. H. (1984) Synthesized warning messages: effects of an alerting cue in single- and multiple-function voice synthesis systems, *Human Factors*, **26**(2), 185–95.

HELLIER, E. J. and EDWORTHY, J. (1989) Quantifying the perceived urgency of auditory warnings, *Canadian Acoustics*, **17**(4), 3–11.

HELLIER, E. J., EDWORTHY, J. and DENNIS, I. D. (1993) Improving auditory warning design: quantifying and predicting the effects of different warning parameters on perceived urgency, *Human Factors*, **35**(4), 693–706.

HELLIER, E. J., EDWORTHY, J. and DENNIS, I. D. (1995) A comparison of different techniques for scaling perceived urgency, *Ergonomics*, **38**(4), 693–706.

HELLMAN, R. and ZWICKER, E. (1990) Magnitude scaling: a meaningful method for measuring loudness annoyance? *Proceedings of the 6th Annual Conference of the International Society for Psychophysics*, 123–38. Wurzburg: Society for Psychophysics.

HILTON, M. E. (1993) An overview of recent findings on alcohol beverage warning labels, *Journal of Public Policy and Marketing*, **12**(1), 1–9.

HOGE, H., SCHICK, A., KUWANO, S., NAMBA, S., BOCK, M. and LAZARUS, H. (1988) Are there invariants of sound interpretation? The case of danger signals, in *Noise as a Public Health Problem*, Stockholm: Swedish Council for Building Research.

HUGHES, P. K. and COLE, B. L. (1986) Can the conspicuity of objects be predicted from laboratory experiments?, *Ergonomics*, **29**, 1097–111.

IRVING, J. (1993) Interior space, in IRVING, J. *Trying to Save Piggy Sneed*, London: Black Swan.

ISO 3744 (1981) Sound Power Levels of Noise Sources: Engineering Methods for free-field conditions over a reflecting plane, Geneva: International Organisation for Standardisation.

ISO 3864 (1984) *Safety Colours and Safety Signs*, Geneva: International Organization for Standardisation.

ISO 7001 (1990) *Public Information Symbols*, Geneva: International Organization for Standardisation.

ISO 9186 (1989) *Procedures for the Development and Testing of Public Information Symbols*, Geneva: International Organization for Standardisation.

ISO/S 9703-2 (in press) Alarm signals for anaesthesia and respiratory care Part 2: Specification for auditory alarm signals.

JACOBS, R. J., JOHNSTON, A. W. and COLE, B. L. (1975) The visibility of alphabetic and symbolic traffic signs, *Australian Road Research*, **5**(7), 68–86.

JAMES, S. H. and JAMES, M. R. (1989) The effects of warning format and keyboard layout on reaction times to auditory warnings, *Proceedings of the Institute of Acoustics*, **11**(5), 25–42.

JANDER, H. F. and VOLK, W. N. (1934) Effectiveness of various highway signs, *Highway Research Board Proceedings*, **14**, 442–7.

JAYNES, L. S. and BOLES, D. B. (1990) The effect of symbols on warning compliance, *Proceedings of the Human Factors Society 34th Annual Meeting*, 984–7, Santa Monica: Human Factors Society.

KAMIN, L. J. (1969) Predictability, surprise, attention and conditioning, in CAMPBELL, B. A. and CHURCH, R. M. (Eds), *Punishment and Aversive Behavior*, New York: Appleton-Century-Crofts.

KERR, J. H. (1985) Warning Devices, *British Journal of Anaesthesia*, **57**, 696–708.

KERR, J. H. and HAYES, B. (1983) An 'alarming' situation in the intensive therapy unit, *Intensive Care Medicine*, **9**, 103.

KING, L. E. (1971) A laboratory comparison of symbol and word roadway signs, *Traffic Engineering and Control*, **12**, 518–20.

KING, L. E. (1975) Recognition of symbol and word traffic signs, *Journal of Safety Research*, **7**, 80–4.

KLINE, D. W. and FUCHS, P. (1993) The visibility of symbolic highway signs can be increased among drivers of all ages, *Human Factors*, **35**, 25–34.

KLINE, P. B., BRAUN, C. C., PETERSON, N. and SILVER, N. C. (1993) The impact of colour on warnings research, *Proceedings of the 37th Meeting of the Human Factors and Ergonomics Association*, 940–4. Santa Monica: Human Factors and Ergonomics Association.

KLINE, T. J. B., GHALI, L. M., KLINE, D. W. and BROWN, S. (1990) Visibility distance of highway signs among young, middle-aged, and older observers: Icons are better than text, *Human Factors*, **32**, 609–19.

KUWANO, S. and NAMBA, S. (1990) Continuous judgement of loudness and annoyance, *Proceedings of the 6th Annual Meeting of the International Society for Psychophysics*. Wurzburg: Society for Psychophysics.

LAROCHE, C., TRAN QUOC, H., HETU, R. and McDUFF, S. (1991) 'Detectsound': a computerised model for predicting the detectability of warning signals in noisy environments, *Applied Acoustics*, **33**(3), 193–214.

LAUGHERY, K. R., VAUBEL, K. P., YOUNG, S. L., BRELSFORD, J. W. and ROWE, A. L. (1993a) Explicitness of consequence information in warnings, *Safety Science*, **16**(5/6), 597–614.

LAUGHERY, K. R., YOUNG, S. L., VAUBEL, K. P. and BRELSFORD, J. W. (1993b) The noticeability of warnings on alcoholic beverage containers, *Journal of Public Policy and Marketing*, **12**(1), 38–56.

LAUGHERY, K. R., ROWE-HALBERT, A. L., YOUNG, S. L., VAUBEL, K. P. and LAUX, L. (1991) Effects of explicitness in conveying severity information in product warnings, *Proceedings of the 35th Annual Meeting of the Human Factors Society*, 499–504, Santa Monica: Human Factors Society.

LAZARUS, H. and HOGE, H. (1986) Industrial safety: acoustic signals for danger situations in factories, *Applied Ergonomics*, **17**(1), 41–6.

LEHTO, M. R. and PAPASTAVROU, J. D. (1993) Models of the warning process: important implications towards effectiveness, *Safety Science*, **16**(5/6), 569–96.

LEHTO, M. R. and MILLER, J. M. (1986) *Warnings: Volume 1: Fundamentals, Design and Evaluation Methodologies*, Ann Arbor, Michnigan: Fuller Technical Publications.

LEONARD, S. D., KARNES, E. W. and SCHNEIDER, T. (1988) Scale values for warning symbols and words, in AGHAZADEH, F. (Ed.), *Trends in Ergonomics/ Human Factors V*. Amsterdam: Elsevier.

LEONARD, S. D., HILL, G. W., IV, and KARNES, E. W. (1989) Risk perception and use of warnings, *Proceedings of the 33rd Annual Meeting of the Human Factors Society*, 550–4, Santa Monica: Human Factors Society.

LERNER, N. D. and COLLINS, B. L. (1983) Symbol sign understandability when visibility is poor, *Proceedings of the Human Factors Society 27th Annual Meeting*, 944–6, Santa Monica: Human Factors Society.

LETOURNEAU, J. E., DENIS, R. and LONDORI, D. (1986) Influence of auditory or visual warning on visual reaction time with variations of subjects' alertness, *Perceptual and Motor Skills*, **62**, 667–74.

LEY, P. (1988) *Communicating with Patients*, London: Chapman and Hall.

LICKLIDER, J. C. R. (1956) Auditory frequency analysis, in CHERRY, C. (Ed.), *Information Theory*, New York: Academic Press.

LOWER, M. C. and WHEELER, P. D. (1985) Specifying the sound levels for auditory warnings in noisy environments, in BROWN, I. D. *et al.* (Eds), *Ergonomics International '85*, 226–8. London: Taylor and Francis.

LOWER, M. C., PATTERSON, R. D., ROOD, G. M., EDWORTHY, J., SHAILER, M. J., MILROY, R., CHILLERY, J. and WHEELER, P. D. (1986) The design and production of auditory warnings for helicopters. 1: the Sea King. Research report No. AC527A, Institute of Sound and Vibration, Southampton, UK.

LOXLEY, S. L. (1991) An investigation of subjective interpretations of auditory stimuli for the design of trend monitoring sounds. Unpublished MPhil thesis, University of Plymouth, UK.

LUCE, P. A., FEUSTEL, T. C. and PISONI, D. B. (1983) Capacity demands in short-term memory for synthetic and natural speech, *Human Factors*, **25**(1), 17–32.

LUNNEY, D., MORRISON, R. C., CETERA, M. M., HARTNESS, R. V., MILLS, R. T., SALT, A. D. and SOWELL, D. C. (1983) A microcomputer-based laboratory aid for visually impaired students, *IEEE Micro*, **3**(4).

MADDUX, J. E. and ROGERS, R. W. (1983) Protection motivation and self-efficacy: a revised theory of fear appeals and attitude change, *Journal of Experimental Social Psychology*, **19**, 469–79.

MALOUFF, J., SCHUTTE, N, FROHARDT, M., DEMING, W. and MANTELLI, D. (1992) Preventing smoking: evaluating the potential effectiveness of cigarette warnings, *The Journal of Psychology*, **126**(4), 371–83.

MANSUR, D. L., BLATTNER, M. M. and JOY, K. I. (1985) Soundgraphs: a numerical data analysis method for the blind, *Proceedings of the 18th Hawaii Conference on System Sciences*, 163–74. Honolulu, Hawaii: IEEE Computer Society Press.

MAYER, D. L. and LAUX, L. F. (1989) Recognizability and effectiveness of warning symbols and pictorials, *Proceedings of the Human Factors Society 33rd Annual Meeting*, 984–8, Santa Monica: Human Factors Society.

McCARTHY, R. L., ROBINSON, J. N., FINNEGAN, J. P. and TAYLOR, R. K. (1982) Warnings on consumer products: objective criteria for their use, *Proceedings of the 26th Human Factors Society*, 98–102, Santa Monica: Human Factors Society.

McCARTHY, R. L., AYRES, T. J. and WOOD, C. T. (1995) Risk and effectiveness criteria for use of on-product warnings, *Ergonomics*, **38**(11), 2164–75.

McINTYRE, J. W. R. (1985) Ergonomics: anaesthetists' use of auditory alarms in the operating room, *International Journal of Clinical Monitoring and Computing*, **2**, 47–55.

McINTYRE, J. W. R. and STANFORD, L. M. (1985) Ergonomics and anaesthesia: auditory alarm signals in the operating room, in DROH, R., ERDMANN, W. and SPINTGE, R. (Eds), *Anaesthesia: Innovations in Management*, 87–92, Springer-Verlag.

MEREDITH, C. S. and EDWORTHY, J. (1994) Sources of confusion in intensive therapy unit alarms, in STANTON, N. A. (Ed.), *Human Factors in Alarm Design*, 207–20. London: Taylor and Francis.

MEYER, L. B. (1956) *The Language of Music*, Chicago: Chicago University Press.

MILLER, G. A. (1956) The magic number seven plus or minus two *Psychological Review*, **63**, 81–97.

MODLEY, R. (1966) Graphic symbols for world-wide communication, in KEPES, G. (Ed.), *Sign, Image and Symbol*, 108–125, London: Studio Vista.

MODLEY, R. (1976) *Handbook of Pictorial Symbols*, New York: Dover.

MOMTAHAN, K. L. (1990) 'Mapping of psychoacoustic parameters to the perceived urgency of auditory warning signals', unpublished Master's thesis, Carleton University, Ottawa, Ontario, Canada.

MOMTAHAN, K. L. and TANSLEY, B. W. (1989) 'An ergonomic analysis of the auditory alarm signals in the operating room and the recovery room', presentation at the Annual Conference of the Canadian Acoustical Association, Halifax, Nova Scotia.

MOMTAHAN, K. L., TANSLEY, B. W. and HETU, R. (1993) Audibility and identification of auditory alarms in the operating room and intensive care unit, *Ergonomics*, **36**, 1159–76.

NEWSTEAD, S. N. and COLLIS, J. (1987) Context and the interpretation of quantifiers of frequency, *Ergonomics*, **30**(10), 1447–62.

NOSULENKO, V. N. (1990) Problems of ecological psychoacoustics. *Proceedings of the 6th Meeting of the International Society for Psychophysics*, 135–8. Wurzburg: Society for Psychophysics.

NOWLIS, V. (1965) Research with the mood adjective check list, in TOMKINS, S. and IZARD, C. (Eds), *Affect, Cognition and Personality*, New York: Springer-Verlag.

O'CARROLL, T. M. (1986) Survey of alarms in an intensive care unit, *Anaesthesia*, **41**, 742–4.

OTSUBO, S. M. (1988) A behavioural study of warning labels for consumer products: perceived danger and use of pictographs, *Proceedings of the 32nd Human Factors Society*, 536–40, Santa Monica: Human Factors Society.

PAIVIO, A. (1986) *Mental Representations: A Dual Coding Approach*, New York: Oxford University Press.

PATE-CORNELL, M. E. (1986) Warning systems in risk management, *Risk analysis*, **6**(2), 223–34.

PATTERSON, R. D. (1974) Auditory filter shape, *Journal of the Acoustical Society of America*, **55**, 802–9.

PATTERSON, R. D. (1976) Auditory filter shapes derived with noise stimuli, *Journal of the Acoustical Society of America*, **59**, 640–54.

PATTERSON, R. D. (1982) Guidelines for auditory warnings systems on civil aircraft, Civil Aviation Authority paper 82017. London: Civil Aviation Authority.

PATTERSON, R. D. and MILROY, R. (1980) Auditory warnings on civil aircraft: the learning and retention of warnings. Civil Aviation Authority contract report 7D/S/ 0142 R3.

PATTERSON, R. D. and MOORE, B. C. J. (1986) Auditory filters and excitation patterns, in MOORE, B. C. J. (Ed.), *Frequency Selectivity and Hearing*, London: Academic Press.

PATTERSON, R. D. and NIMMO-SMITH, I. (1980) Off-frequency listening and auditory filter asymmetry, *Journal of the Acoustical Society of America*, **67**, 229–45.

PATTERSON, R. D., EDWORTHY, J., SHAILER, M. J., LOWER, M. C. and WHEELER, P. D. (1986) Alarm sounds for medical equipment in intensive care areas and operating theatres. Report no. AC598, Southampton: UK, Institute of Sound and Vibration Research.

PETERS, G. A. (1984) A challenge to the safety profession, *Professional Safety*, **29**, 46–50.

POLZELLA, D. J., GRAVELLE, M. D. and KLAUER, K. M. (1992) Perceived effectiveness of danger signs: a multivariate analysis, *Proceedings of the 36th Annual Meeting of the Human Factors Society*, 931–4, Santa Monica: Human Factors Society.

RACICOT, B. M. and WOGALTER, M. S. (1992) Warning compliance: effects of a video warning sign and modelling on behaviour, *Proceedings of the Human Factors Society 36th Annual Meeting*, 608–10, Santa Barbara: Human Factors Society.

RILEY, M. W., COCHRAN, D. J. and BALLARD, J. L. (1982) An investigation of preferred shapes for warning labels, *Human Factors*, **24**, 737–42.

ROBINSON, C. P. and EBERTS, R. E. (1987) Comparison of speech and pictorial displays in a cockpit environment, *Human Factors*, **29**(1), 31–44.

ROOD, G. M. (1989) Auditory warnings for fixed and rotary wing aircraft, *Proceedings of the Institute of Acoustics*, **11**(5), 59–72.

ROOD, G. M., CHILLERY, J. A. and COLLISTER, J. B. (1985) Requirements and applications of auditory wanrings to military helicopters, in BROWN, I. D. *et al.* (Eds), *Ergonomics International '85*, 169–71, London: Taylor and Francis.

ROTHSTEIN, P. R. (1985) Designing warnings to be read and remembered, *Proceedings of the 29th Annual Meeting of the Human Factors Society*, 684–8, Santa Monica: Human Factors Society.

SCHOUTEN, J. F. (1940) The perception of timbre, *Reports on the 6th International Congress of Acoustics*, **1**, GP-6-2. Tokyo, Japan.

SCHREIBER, P. J. and SCHREIBER, J. (1989) Structured alarm systems for the operating room, *Journal of Clinical Monitoring*, **5**(3), 201–4.

SCHWARZER, R. (1992) Self-efficacy in the adoption and maintenance of health behaviours: theoretical approaches and a new model, in SCHWARZER, R. (Ed.), *Self-Efficacy: Thought Control of Action*. Washington: Hemisphere.

SCHWARTZ, D. R., DE PONTBRIAND, R. J. and LAUGHERY, K. R. (1983) The impact of product hazard information on consumer buying decisions: a policy-capturing approach, *Proceedings of the 27th Annual Meeting of the Human Factors Society*, 955–7, Santa Monica: Human Factors Society.

SENNER, W. M. (Ed.) (1989) *The Origins of Writing*, Lincoln: University of Nebraska Press.

SILVER, N. C. and BRAUN, C. C. (1993) Perceived readability of warning labels with varied font sizes and styles, *Safety Science*, **16**(5/6), 615–26.

SILVER, N. C. and WOGALTER, M. S. (1991) Strength and understanding of signal words by elementary and middle school students, *Proceedings of the 35th Annual Meeting of the Human Factors Society*, 590–4, Santa Monica: Human Factors Society.

SILVER, N. C., LEONARD, D. C., PONSI, K. A. and WOGALTER, M. S. (1991) Warnings and purchase intentions for pest-control products, *Forensic Reports*, **4**, 17–33.

SIMPSON. T. (1991) Attensons and monitors for the European fighter aircraft. British Aerospace Sowerby Research Centre report No. JS11593. Sowerby, UK.

SIMPSON, C. A. and HART. S. G. (1977) Required attention for synthesized speech perception for two levels of linguistic redundancy, *Journal of the Acoustical Society of America*, **61**, suppl. 1, S7/D3.

SIMPSON, C. A. and MARCHIONDA-FROST, K. (1984) Synthesized speech rate and pitch effects on intelligibility of warning messages for pilots, *Human Factors*, **26**(5), 509–17.

SIMPSON, C. A. and WILLIAMS, D. H. (1980) Response time effects of alerting tone and semantic context for synthesized voice cockpit warnings, *Human Factors*, **22**(3), 319–30.

SIMPSON, C. A., MCCAULEY, M. E., ROLAND, E. F., RUTH, J. C. and WILLIGES, B. H. (1987) Speech controls and displays, in SALVENDY, G.(Ed.), *Handbook of Human Factors*, 1490–525, New York: Wiley.

SLOWIACEK, L. M. and NUSBAUM, H. C. (1985) Effects of speech rate and pitch contour on the perception of synthetic speech, *Human Factors*, **27**(6), 701–12.

SOLOMON, L. N. (1959a) Semantic approach to the perception of complex sounds, *Journal of the Acoustical Society of America*, **30**, 421–5.

SOLOMON, L. N. (1959b) Search for physical correlates to psychological dimensions of sounds, *Journal of the Acoustical Society of America*, **31**, 292–497.

SOLOMON, L. N. (1959c) Semantic reactions to systematically varied sounds, *Journal of the Acoustical Society of America*, **31**, 986–90.

SORKIN, R. D. (1982) Design of auditory and tactile displays, in SALVENDY, G. (Ed.), *Handbook of Human Factors*, 549–76, New York: Wiley.

SORKIN, R. D., KANTOWITZ, B. H. and KANTOWITZ, S. C. (1988) Likelihood alarm displays, *Human Factors*, **30**(4), 445–9.

STANTON, N. A. and EDWORTHY, J. (1994) Towards a methodology for the design of representational auditory alarms, in ROBERTSON, S. A. (Ed.), *Contemporary Ergonomics 1994*, 360–5. London: Taylor & Francis.

STANFORD, L. M., MCINTYRE, J. W. R. and HOGAN, J. T. (1985) Audible alarm signals for anaesthesia monitoring equipment, *International Journal of Clinical Monitoring and Computing*, **1**, 251–6.

STANFORD, L. M., MCINTYRE, J. W. R., NELSON, T. M. and HOGAN, J. T. (1988) Affective responses to commercial and experimental auditory alarm signals for anaesthesia delivery and physiological monitoring equipment, *International Journal of Clinical Monitoring and Computing*, **5**, 111–8.

STEVENS, S. S. (1957) On the psychophysical law, *Psychological Review*, **64**, 153–81.

STEVENS, S. S. and GALANTER, E. (1957) Ratio scales and category for a dozen perceptual continua, *Journal of Experimental Psychology*, **54**, 377–411.

STRAWBRIDGE, J. A. (1986) The influence of position, highlighting, and imbedding on warning effectiveness, *Proceedings of the 30th Annual Meeting of the Human Factors Society*, 716–20, Santa Monica: Human Factors Society.

SUTTON, S. (1987) Social-psychological approaches to understanding addictive behaviours: attitude-behaviour and decision-making models, *British Journal of Addiction*, **82**, 355–70.

SUTTON, S. R. and EISER, J. R. (1990) The decision to wear a seat-belt: the role of cognitive factors, *Psychology and Health*, **4**, 111–24.

THORNING, A. G. and ABLETT, R. M. (1985) Auditory warning systems on commercial transport, in BROWN, I. D. *et al.* (Eds), *Ergonomics International '85*, 166–8. London: Taylor and Francis.

URSIC, M. (1984) The impact of safety warnings on perception and memory, *Human Factors*, **26**(6), 677–82.

WALKER, R. E., NICOLAY, R. C. and STEARNS, C. R. (1965) Comparative accuracy of recognizing American and international road signs, *Journal of Applied Psychology*, **49**, 322–5.

WARREN, W. H. and VERBRUGGE, R. R. (1984) Auditory perception of breaking and bouncing events: a case study in ecological acoustics, *Journal of Experimental Psychology: Human Perception and Performance*, **10**(5), 704–12.

WEBSTER, J. C., WOODHOUSE, M. M. and CARPENTER, A. S. (1973) Perceptual confusions between four-dimensional sounds, *Journal of the Acoustical Society of America*, **53**, 448–56.

WEINSTEIN, N. D. (1993) Testing four competing theories of health-protective behaviour, *Psychology and Health*, **3**, 259–85.

WEINSTEIN, A. S., TWERSKI, A. D., PIEHLER, H. R. and DONAHER, W. A. (1978) *Products Liability and the Reasonably Safe Product*, New York: Wiley & Sons.

WELKER, R. L. (1982) Abstraction of themes from melodic variations, *Journal of Experimental Psychology: Human Perception and Performance*, **8**(3), 435–47.

WESTINGHOUSE ELECTRIC CORPORATION (1981) *Westinghouse Product Safety Label Handbook*, Trafford, PA: Author.

WICKENS, C. D. (1992) *Engineering Psychology and Human Performance*, New York: HarperCollins.

WILKINS, P. and MARTIN, A. M. (1987) Hearing protection and warning sounds in industry: a review, *Applied Acoustics*, **21**, 267–93.

WITTE, K., STOKOLS, D., ITUARTE, P. and SCHNEIDER, M. (1993) Testing the health belief model in a field study to promote bicycle safety helmets, *Communication Research*, **20**(4), 564–86.

WOGALTER, M. S. and SILVER, N. C. (1990) Arousal strength of signal words, *Forensic Reports*, **3**, 407–20

WOGALTER, M. S. and SILVER, N. C. (1995) Warning signal words: connoted strength and understandability by children, elders and non-native english speakers, *Ergonomics*, **38**(11), 2188–206.

WOGALTER, M. S. and YOUNG, S. L. (1991) Behavioural compliance to voice and print warnings, *Ergonomics*, **34**, 78–89.

WOGALTER, M. S. and YOUNG, S. L. (1994) The effect of alternative product-label design on warning compliance, *Applied Ergonomics*, **25**(1), 53–7.

WOGALTER, M. S., ALLISON, S. P. and McKENNA, N. A. (1989) Effects of cost and social influence on warning compliance, *Human Factors*, **31**(2), 133–40.

WOGALTER, M. S., BARLOW, T. and MURPHY, S. A. (1995) Compliance to owner's manual warnings: influence of familiarity and the placement of a supplemental directive, *Ergonomics*, **38**, 1081–91.

WOGALTER, M. S., BRELSFORD, J. W., DESAULNIERS, D. R. and LAUGHERY, K. R. (1991) Consumer product warnings: the role of hazard perception, *Journal of Safety Research*, **22**, 71–82.

WOGALTER, M. S., DESAULNIERS, D. R. and BRELSFORD, J. W. (1986) Perceptions of consumer products: hazardousness and warning expectations, *Proceedings of the 30th Annual Meeting of the Human Factors Society*, 1197–2001, Santa Monica: Human Factors Society.

WOGALTER, M. S., GODFREY, S. S., FONTENELLE, G. A., DESAULNIERS, D. R., ROTHSTEIN, P. R. and LAUGHERY, K. R. (1987) Effectiveness of warnings, *Human Factors*, **29**(5), 599–612.

WOGALTER, M. S., JARRARD, S. W. and SIMPSON, S. N. (1992) Effects of warning signal words on consumer-product hazard perceptions, *Proceedings of the 36th Annual Meeting of the Human Factors Society*, 935–9. Santa Monica:. Human Factors Society.

WOGALTER, M. S., KALSHER, M. J. and RACICOT, B. M. (1993) Behavioural compliance with warnings: effects of voice, context and location, *Safety Science*, **16**(5/6), 637–54.

YOUNG, S. L. (1991) Increasing the noticeability of warnings: effects of pictorial, colour, signal icon and border, *Proceedings of the 35th Annual Meeting of the Human Factors Society*, 580–4, Santa Monica: Human Factors Society.

YOUNG, S. L. and WOGALTER, M. S. (1990) Comprehension and memory of instructions manual warnings: Conspicuous print and pictorial icons, *Human Factors*, **32**, 637–49.

YOUNG, S. L., LAUGHERY, K. R. and BELL, M. (1992) Effects of two type density characteristics on the legibility of print, *Proceedings of the 36th Annual Meeting of the Human Factors Society*, Santa Monica: Human Factors Society.

ZEDECK, S. and KAFRY, D. (1977) Capturing rater policies for processing evaluation data, *Organisational Behaviour and Human Performance*, **18**, 269–74.

ZWAGA, H. J. (1989) Comprehensibility estimates of public information symbols; their validity and use, *Proceedings of the Human Factors Society 33rd Annual Meeting*, 979–83, Santa Monica: Human Factors Society.

ZWAGA, H. J. G., HOONHOUT, H. C. M. and VAN GEMERDEN, B. (1991) The systematic development of a set of pictographic symbols for warnings and product information, *Proceedings of the Ergonomics Society 1991 Annual Conference*, 233–8, Santa Monica: Human Factors Society.

ZWICKER, E. and SCHARF, B. (1965) A model of loudness summation, *Psychological Review*, **72**, 3–26.

Index

Permissions

Chapter 2
Figure 2.5 and Table 2.11 reprinted from Laughery, K. R., Vaubel, K. P., Young, S. L., Brelsford, J. W. and Rowe, A. L. (1993) Explicitness of consequence information in warnings, *Safety Science*, **16** (5/6) 597-614. With kind permission of Elsevier Science - NL, Sara Burgerhartstraat 25, 1055 KV Amsterdam, The Netherlands.

Table 2.7 reprinted from Wogalter, M. S., Kalsher, M. J. & Racicot, B. M. (1993) Behavioral compliance with warnings: effects of voice, context and location, *Safety Science*, **16** (5/6), 637-654. With kind permission of Elsevier Science - NL, Sara Burgerhartstraat 25, 1055 KV Amsterdam, The Netherlands.

Chapter 3
Figure 3.4 reprinted from Wogalter, M. S. & Young, S. L. (1994) The effects of alternative product-label design on warning compliance, *Applied Ergonomics*, **25** (1), 53-7. With kind permission of Elsevier Science Ltd, The Boulevard, Langford Lane, Kidlington, OX5 1GB, UK.

Figure 3.1 Reproduced with permission from: Modley, R. (1976) *Handbook of pictorial symbols*. New York: Dover.

Figure 3.2 Reproduced from: Bliss, C. K. (1968) *Semantography-Blissymbolics*. Sydney: Semantography-Blissymbolics Publications.

Chapter 4
Figure 4.1 and Tables 4.1, 4.2 reprinted from McIntyre, J. W. R. (1985) Ergonomics: anaesthetists' use of auditory alarms in the operating room, *International Journal of Clinical Monitoring and Computing*, **2**, 47-55. With permission from Kluwer Academic Publishers.

Table 4.3 published by kind permission of Springer-Verlag, Berlin. From McIntyre, J.W. R. & Stanford, L. M. (1985) Ergonomics and Anaesthesia: auditory alarms in the operating room. In Droh, R., Erdmann, W. & Spintge, R. (Eds) *Anaesthesia: Innovations in Management*, pp. 87-92.

Table 4.4 reprinted from Kerr, J H (1985) Warning Devices, *British Journal of Anaesthesia*, **57**, 696-708. With kind permission from Professor G. Smith, Editor, *British Journal of Anaesthesia*, Leicester Royal Infirmary.

Chapter 5
Figure 5.3 reprinted from Laroche, C., Tran Quoc, H., Hetu, R. & McDuff, S. (1991) 'Detectsound': a computerized model for predicting the detectability of

warning signals in noisy environments, *Applied Acoustics*, **33** (3), 193-214. With kind permission from Elsevier Science Ltd, The Boulevard, Langford Lane, Kidlington, OX5 1GB, UK.

Figures 5.1 and 5.2 reprinted from Civil Aviation Authority paper 82017, Patterson, R. D. (1982) Guidelines for auditory warning systems on civil aircraft. With kind permission from Mr A. Doyle, Civil Aviation Authority Library & Information Centre, Aviation House, Gatwick Airport South, West Sussex RH6 0YR

Table 5.3 reprinted from Stanford, L M, McIntyre, J. W. R., Nelson, T. M. & Hogan, J. T. (1988) Affective responses to commercial and experimental auditory alarm signals for anaesthesia delivery and physiological monitoring equipment, *International Journal of Clinical Monitoring and Computing*, **5**, 111-18. With permission from Kluwer Academic Publishers.

Table 5.9 reprinted from Edworthy, J. (1994) The design and implementation of nonverbal auditory warnings, *Applied Ergonomics*, **25**, 202-10. With kind permission from Elsevier Science Ltd, The Boulevard, Langford Lane, Kidlington, OX5 1GB, UK.

Tables 5.14 and 5.15 reprinted with kind permission from Loxley, S. L., 16a Bare Avenue, Morecombe, Lancs LA4 6BE.

Chapter 6
Table 6.1 reprinted from Cowley, C. K. and Jones, D. M. (1992) Synthesized or digitized? A guide to the use of computer speech, *Applied Ergonomics*, **23** (3), 172-6. With kind permission from Elsevier Science Ltd, The Boulevard, Langford Lane, Kidlington, OX5 1GB, UK.

The following figures and tables are reprinted with the kind permission of the Human Factors and Ergonomics Society, Box 1369, Santa Monica, California 90406-1369, USA:

Figure 2.3 from Braun, C. C., Sansing, L., Kennedy, R. S. and Silver, N. C. (1994) Signal word and color specifications for product warnings: An isoperformance application, *Proceedings of the Human Factors and Ergonomics Society 38th Annual Meeting*, 1104-1108. Copyright 1994 Human Factors and Ergonomics Society. All rights reserved.

Table 2.5 from Dejoy, D. M. (1989) Consumer product warnings: review and analysis of effectiveness research, *Proceedings of the 33rd Annual Meeting of the Human Factors Society*, 936-9. Copyright 1989 Human Factors and Ergonomics Society. All rights reserved.

Figure 2.8 from Wogalter, M. S., Allison, S. P. and McKenna, N. A. (1989) Effects of cost and social influence on warning compliance, *Human Factors*, **31** (2), 133-40. Copyright 1989 Human Factors and Ergonomics Society. All rights reserved.

Table 5.6 from Edworthy, J., Loxley, S. & Dennis, I. (1991) Improving auditory warning design: relationship between warning sound parameters and perceived urgency, *Human Factors*, **33** (2), 205-31. Copyright 1991 Human Factors and Ergonomics Society. All rights reserved.

Table 6.2 from Simpson, C. A. and Williams, D. H. (1980) Response time effects of altering tone and semantic context for synthesized voice cockpit warnings, *Human Factors*, **22** (3), 319-30. Copyright 1980 Human Factors and Ergonomics Society. All rights reserved.

Figure 6.1 from Slowiacek, L. M. & Nussbaum, H. C. (1985) Effects of speech rate and pitch contour on the perception of synthetic speech, *Human Factors*, **27**, 701-12. Copyright 1985 Human Factors and Ergonomics Society. All rights reserved.

Table 6.3 from Sorkin, R. D., Kantoqitz, B. H. and Kantowitz, S. C. (1988) Likelihood alarm displays, *Human Factors*, **30** (4), 445-59. Copyright 1988 Human Factors and Ergonomics Society. All rights reserved.